Making of a Soil Scientist

CABI is a trading name of CAB International

CABI
Nosworthy Way
Wallingford
Oxfordshire OX10 8DE
UK

Tel: +44 (0)1491 832111
E-mail: info@cabi.org
Website: www.cabi.org

CABI
200 Portland Street
Boston
MA 02114
USA

T: +1 (617)682-9015
E-mail: cabi-nao@cabi.org

A catalogue record for this book is available from the British Library, London, UK.

ISBN-13: 9781836990130 (hardback)
9781836990147 (ePDF)
9781836990154 (ePub)

DOI: 10.1079/9781836990154.0000

Commissioning Editor: David Hemming
Editorial Assistant: Helen Elliott
Production Editor: Shankari Wilford

Typeset by Exeter Premedia Services Pvt Ltd, Chennai, India
Printed and bound in the UK by CPI Group (UK) Ltd, Croydon, CR0 4YY

Foreword

As a young boy growing up in rural Tennessee, I learned from my father how to find where the most fertile soil was located on our family's farm. Many years later I learned that – at the same time as I was digging into the ground in America – on the other side of the world, in rural north-western India, another young boy was doing his own digging on his family's farm. It was on that small farm that a young and curious Rattan Lal embarked on his path to become the world's undisputed foremost expert on soil science.

As you will read in this autobiography, it was not just his curiosity but also an unshakeable determination that drove Dr Lal to overcome the challenges of life as a refugee in rural India. As subsistence farmers, Rattan Lal and his family were acutely aware of how the whims of Mother Nature could affect their lives. With droughts and heatwaves already common in their region, one of the most important actions his family could take to ensure the survival of their crops was to care for the soil in which those crops grew. As Dr Lal embarked on his career, it became clear to him that what was known by generations of subsistence farmers around the world was not well understood – and was often even ignored – across the global food system.

Dr Lal became a pioneer in advocating for a soil-centric approach to agriculture – one that supports increased crop yields, protects natural resources, and sequesters carbon. It is the connection between the soils of our planet and the global climate crisis that led me to Dr Lal's work. It was through his work that I finally understood why the rich soil we would dig up on the farm was so important ... and so black: it was the carbon. Dr Lal's scientific research has informed my own book *Our Choice: A Plan to Solve the Climate Crisis* and is central to the conference on regenerative agriculture that I hold at my farm each year, The Climate Underground.

Our food system accounts for 30% of global greenhouse gas emissions: 15% of the global total is from crop and livestock production within the farm gate, and the remainder is from supply chains and land use change. Dr Lal's research shows that if we were to shift away from the conventional agricultural practices and toward a regenerative approach that prioritizes the health of the soil, the technical potential of carbon sequestration in soils and vegetation this century is 156 parts per million of CO_2.

The potential for regenerative agriculture to be part of the solution to the climate crisis is significant.

Dr Lal's research has also shown that by restoring the carbon content of soil, you can significantly increase and sustain the quality and quantity of agricultural yields. This is particularly relevant in tropical regions of developing countries where soil degradation has fed a vicious cycle of hunger and poverty.

By restoring soil's carbon content, farmers are not only able to help sustain their livelihoods, but they can also build their resilience to the impacts of the climate crisis we are already experiencing and become part of the solution to the crisis by storing carbon.

I have seen the benefits of this approach first-hand – in the soils of the same family farm that I dug into as a kid. These days, the digging we do is a little different. I converted my family farm to regenerative agriculture practices about ten years ago. We keep tilling to a minimum, integrate agroforestry, and focus on rotational livestock grazing to turn over the land. When we dig deep, it is to take soil samples to measure the carbon content of the soil.

Dr Lal has been a frequent visitor to the farm, where we discuss his latest research and he advises our team on the most effective practices for carbon sequestration. When we are out in the fields, I can see the spark of curiosity light up in Rattan's eyes as if he were a young boy back on his family's farm.

His curiosity inspires my own. His research offers much-needed hope in the face of the world's most pressing global challenges: hunger, poverty, and the climate crisis.

And so it is my hope that this autobiography will inspire you. Dr Lal's story and lessons are both powerful and practical. It is my sincere hope that everyone who reads this book will take Rattan's story to heart and take action into your own hands. Look to your own backyards and find a way to deploy natural climate solutions where you live.

We have the power to reshape our world for the better. Now is the time to dig into this challenge and get our hands dirty.

Al Gore,
former US Vice-President

Preface

This book is the autobiography of a soil scientist who had the great privilege and opportunities to grow up on a subsistence farm in a rural community of north-western India. In Sanskrit, the word Rattan is a variation of the word 'Ratna' (रतन), which means jewel or gem. For example, Emperor Akbar had 'Nav Ratna' or nine gems in his Durbar. Lal (Persian: لعل, Hindi: लाल, Bengali: লাল, Urdu: لال, Pashto: Lāl) is the name for a male child. It is an Indo-Iranian surname and a given name, which means darling, precious, or beloved. Lal also means garnet or ruby in Persian, ruby in Pashto, and red in Hindustani and Bengali. Rattan Lal's family migrated from west Punjab, now in Pakistan, at the time of the partition of India in 1947 when he was about five years old. The family's gotra (sub-caste of Brahmin) is Bhakri. Thus, his full name is Rattan Lal Bhakri.

Rattan had the good fortune and God's blessing to be able to transform challenges of a refugee growing up on a small farm into opportunities for professional advancement. He was native of a region in which Lahore (along with Balkh and Bukhara in the neighbouring Uzbekistan to the north) was the famous city of the ancient times. Furthermore, he had a family background which provided an opportunity to learn from synthesis of diverse cultures. He remembers two common phrases about Lahore; one was, 'If you have not seen Lahore, you are not yet born'. Yes, Lahore was among the most famous cities of pre-independence India. The second phrase, among common slogans of the community and one which became the guiding principle of his life, was, 'Jera aithe phaira oh Lahore bhee phaira' ('One who is bad in his own village will also be bad even if he were to shift to Lahore'). However, he was fortunate in having this negative phrase turned around because of the opportunities provided and his rural upbringing. Contrary to what the phrase says, it turned out that he was given a chance to prove that one's background is less important than commitment to do one's best under challenging circumstances. Thus, the fact that he was the best student in the rural village school became the motivational force to work hard and achieve the best position in the class in Ludhiana, Delhi and elsewhere during his education, and in his professional career later at institutions around the world in six continents over six decades.

This book explains Lal's experiences of hardship and challenges in a rural community and village school and the quest to transform these into great advantages and assets by making scientific contributions in the field of soil science and agronomy. Indeed, these experiences made him focused, mission-oriented and targeted at developing research and educational programmes for transforming agriculture/soil management from being problems into a part of the solution, and using these as scientific tools for addressing global issues of the era.

The world of the first three decades of 21st century, when Lal wrote this autobiography, during 2023–2025, was at a crossroads and faced multiple challenges. Ten persistent global challenges of the 21st century in 2025, driven by demands of the ever-growing and increasingly affluent population, were:

1. population of 8.2 billion and increasing at 1.14% each year (7.65 million/month, 92 million/yr);
2. per capita arable land area of 0.186 ha and decreasing to < 0.07 ha for 30 countries by 2025, and some countries in South Asia having per capita arable land of < 0.05 ha;
3. soil degradation affecting 40–50% of the ice-free and desert-free land area of the Earth with direct and adverse impacts on food justice and human rights;
4. per capita renewable fresh water supply of < 1000m^3 for 30 countries and projected to increase to 58 countries (affecting around 4 billion people) by 2050;
5. atmospheric CO_2 concentration of 428.37 ppm on 19 January 2025 and increasing at 0.5%/yr, or ~2 ppmv/yr;
6. energy use of 587 quads (10^{15} BTU/year) or 619.3EJ in 2024, increasing by 3.3%/yr and projected to reach 736 quads by 2040;
7. per capita grain consumption of 300 kg/yr, which may decrease because of pandemics, war and climate change, and access to it that differed widely among countries and geographical regions;
8. food-insecure population of 733 million in 2024 and increasing, and malnourished population of 2–3 billion because of deficiencies of micronutrients, protein and vitamins, and increasing due primarily to food grown on degraded soils;
9. per capita global CO_2 emission of 4.88 Mg CO_2/yr compared with 17.56 Mg CO_2/yr in the USA, and much greater in some oil-exporting nations; and
10. world peace in jeopardy because of the growing scarcity of natural resources, insecurity of food and nutrition, increasing trend in poverty, and aggravating desperation of the growing and increasingly affluent world population.

Despite great progress in agriculture since the 1960s, the world food systems were rather ineffective in addressing major issues, and 'business as usual' was not an option. Limitations of these food systems led to failure at national and global levels and created an urgency in: (a) ending hunger and malnutrition; (b) mitigating human suffering from inadequate, unsafe, unhealthy and less-nutritious diets for people, especially in developing countries; (c) reducing vulnerability of soils to degradation, pollution and contamination; (d) reversing global warming by reducing emissions of greenhouse gases from agroecosystems; (e) improving quality and renewability of water; (f) addressing the so-called drought-flood syndrome; (g) changing soil from being a source into a sink for atmospheric greenhouse gases (especially CO_2); (h) growing carbon on land (soil and trees) as a new crop or commodity that can create another income stream by farming carbon; and (i) promoting peace and social justice for all inhabitants of the planet. Indeed, there were ever-increasing problems with scarcity and renewability of water and its eutrophication, contamination of air, dwindling of biodiversity, denudation of landscape, and degradation of soil which were a multiplier of anthropogenic global warming. Prime soils of the world had also reached the level of 'peak soil'. Some soils had already gone extinct, and others were on the path to becoming endangered because of anthropogenic activities of land misuse and soil mismanagement.

Therefore, Rattan's motivation for writing this autobiography was the strong desire to bring about the transformation of global food systems by narrating the personal story of someone who grew up in a rural community in north-western India. The opportunity to receive a college education guided him to pursue a career in soil sciences. Thus, this book is a narration of his professional life, spanning over six decades and six continents, which has been closely associated with strong professional interests in soil protection, restoration and judicious management of a finite and fragile resource base on which depends all terrestrial life. The basis of Lal's professional interests was the challenges experienced while growing up on a traditional farm where soils were degraded, farm operations were manual, drought and heatwaves were the norm, low agronomic yield of poor nutritional quality was a fact of life, and the need to bring about a paradigm shift was imminent. While Mother Nature can be hard and unforgiving, she turned out to be the best teacher for this budding researcher in soil science. Therefore, the book shows a strong interconnectivity between the childhood experiences of growing up on a small farm with research priorities of a soil scientist during his later career. Because Lal had first-hand experiences in traditional agriculture, as practised on degraded and depleted soils, he focused on promoting the soil-centric approach in his professional career. One of his great advantages was the work experience with the CGIAR system (IITA in Nigeria) for 18 years and with a Land Grant University or LGU (The Ohio State University) for 40 years. He documented that a major part of the solution to these and other global issues was sustainable management of soil and agriculture, and in the protection, restoration and sustainable management of soil health.

Thus, this autobiography narrates the personal journey from childhood on a subsistence farm towards the making of a soil scientist interested in addressing global issues. It explains the basis of the dedication to serve all famers of the world, but especially almost 550 million small landholder farmers, because he considered himself one of them. He regarded their challenges and difficulties as his own and wanted to address these constraints by using the scientific knowledge that he learned with CGIAR and the LGU system.

In addition to strong faith in science, the basis of Lal's cultural belief of *Vasudhavai kutumbkum* ('The world is one family') is traced back to his family heritage and teachings of his grandmother. Indeed, this autobiography was written with the aim of enhancing the prestige of the soil science and agricultural professions so that no farmer is left behind, no person is without adequate food, and no child is without an opportunity to learn and take control of his/her own destiny.

This autobiography also depicts the urgency for transformation of agriculture and soil management into a part of the solution, and shows the pathway of translating soil science into action for addressing global issues in the 21st century.

Rattan Lal
31 March 2025
Columbus, Ohio, USA

Acknowledgements

Writing of the book was made possible because of the inspiration, guidance and support of the family of Rattan Lal, colleagues at the Ohio State University, and friends and co-operators from around the world. I used material from literature and photographs from diverse sources. While it is difficult to list each source and personally thank every individual who contributed, it is important to mention that this volume could not have been completed without the support from numerous listed and unlisted sources, and several acknowledged and unacknowledged contributors.

My special thanks are to two mentors in soil science: Prof. D.R. Bhumbla, who advised me from 1961 to 1963 during my BSc degree at Ludhiana, and encouraged me to major in soil science. Dr Bhumbla also wrote a letter of recommendation to Prof. George S. Taylor and suggested that he advise me for conducting the PhD degree; and Prof. George S. Taylor, who was the academic advisor for my PhD degree, and suggested that I take up a post-doctoral position at the University of Sydney to broaden my outlook in soil science. He visited my research program in Nigeria twice when I was working at the International Institute of Tropical Agriculture (IITA). Prof. Taylor was also responsible for sponsoring me and my family to come to Ohio State. He provided me with the unique opportunity, honour and privilege to succeed him at Ohio State as Professor of Environmental Soil Physics. He was also the grandfather of my children and attended many of their functions at school. For the first couple of years after our arrival in Ohio, Prof. Taylor made sure that we were well settled in our new homeland.

I had professionally rewarding experiences of working at IITA, Ibadan, Nigeria, from December 1969 to July 1987. It was a pleasure working with several Director Generals of IITA, including Dr Herb Albrecht, Dr Bill Gamble and Dr Larry Stifel. Directors of Research at IITA included Dr Rodney Briggs, Dr Dennis Greenland and Dr S.V.S. Shastry. Dr Bede Okigbo was Head of the Farming System Program and Deputy Director General of IITA. I learnt soil sciences by working with Dr Greenland and plant sciences with Dr Bede Okigbo. They were great scientists to work with. Several long-term field experiments on no-till farming, soil erosion measurements and root laboratory were established in co-operation with Doug Cooper and Stan Classen, who were farm managers of IITA. The large watershed management experiment, also called the West Bank experiment, involving different techniques of deforestation, tillage methods, alley cropping and agroforestry systems, was set up in 1977/78 on a 40-hectare watershed with strong support from Doug Cooper and Stan Classen. Soil scientists at IITA during that time included Dr Frank Moorman, Dr B.T. Kang, Dr A.S.R. Juo and Dr A. Ayanaba. Dr Ted Lawson was the agro-climatologist. It was a great team of scientists to work with and gain experiences in managing soils of the humid and sub-humid tropics.

I must thank former 45th US Vice-President Albert Arnold Gore Jr. for providing the Foreword to the book. It has been a great privilege and honor to visit the Carthage Group Farm near Nashville and take part in panel discussions in events on sustainable management of soil health held at his farm. Mr Gore was also the keynote speaker for the 2020 World Food Prize event, which was virtual because of the COVID pandemic.

Several Presidents of The Ohio State University were highly supportive. My sincere thanks are due to President and Mrs Michael Drake, who both attended the Japan Prize ceremony in Tokyo in 2019 and invited me to be the commencement speaker for the 2019 Summer Commencement, and to a football game, along with other dignitaries such as Governor Mike DeWine. President Kristina Johnson approved support of US$ 50,000 per annum for five years as a match to the $1 million donated by me. President Johnson also launched the Carbon Farming project in March 2022. Dr Britt Kirwan, Dr Karen Holebrook and Dr Gordon Gee led delegations to visit India and especially the Punjab Agricultural University (PAU), Ludhiana. The visit to PAU for its 50th anniversary also included Jeff Kaplan and Ted Celeste, who were members of the Board of Trustees of The Ohio State University, and I was a member of the delegation. Several vice-presidents of research at The Ohio State University were supportive of the research on soil carbon sequestration, including Dr Brad Moore, Dr Morley Stone and Dr Peter Mohler. All deans of the colleges of agriculture (and later the College of Food, Agriculture and Environment Sciences) from 1987 onwards supported the soil health and carbon research programme. The centre was created by Dean Dr Bobby Moser, Associate Dean and Director of Research Dr Steve Slack, Dr David Benfield and School of Environment and Natural Resources (SENR) Director Dr Gary W. Mullins. The new name for the centre (CFAES Rattan Lal Center for Carbon Management and Sequestration) was authorized and launched in 2020 by Dean Dr Cathan Kress, Associate Dean Dr Gary Pierzynski, and Director of SENR Dr Jeff Sharp. I and my family greatly appreciated the naming of the centre (Lal Carbon Center) by Dean Dr Kress and Associate Dean Dr Pierzynski. Dr Bruce McPheron supported the Centre when he was Dean of the CFAES and Provost of The Ohio State University. Similarly, I had strong support from all heads of the Department of Agronomy (Dr Berlie Schmidt and Dr Fred Miller), and later the directors of the School of Environment and Natural Resources (e.g. Dr Gary Mullins, Dr Jerry Bigham, Dr J. Sharp and Dr Timothy Haab). I had strong co-operation with the office of the International Program in Agriculture (IPA), especially with Dr David Hansen. Together, we organized several international conferences and planned visits of Ohio State delegations to India. With advice form Dr Moser, Dr Hansen and I nominated (16) academicians, policy-makers, industrialists and philanthropists to receive honorary degrees from The Ohio State University. I also worked with Mr Mike Chrisman, Dr Mark Erbaugh and Dr Luis Canas of the offices of IPA. In the Ohio State Office of International Affairs (OIA), I had great support from Dr Francille Firebaugh, Dr Jerry Ladman, Dr Gil Latz and Dr Kaya Sahin. Dr Firebaugh facilitated the visa for me and my family to the USA. In addition to attending the Japan Prize event, Dr and Mrs Gil Latz also attended several events organized by the Federation of Indian Associations of Central Ohio.

I thank many colleagues and friends from around the world for their co-operation and for providing comments for the book, including Dr Takashi Kosaki, Dr Pete Smith, Dr Bal Ram Singh, Dr Alex McBratney, Dr Joachim Von Braun, Dr Karl-Heinz Feger, Dr Alfred Hartemink, Dr R. Horn and others. I had special co-operation from Dr Bal Ram Singh of the Norwegian University of Life Sciences. He and I were students at the Indian Agricultural Research Institute (IARI) in New Delhi in 1964. While he went to Norway for his postdoc and got settled there, I worked at IITA and The Ohio State University. Dr Bal Ram Singh and I worked together on many projects funded by NORAD in Tanzania, Zambia, Ethiopia and elsewhere. We were co-editors of several books and it was an excellent partnership. He was also responsible for me receiving an honorary degree from NMBU.

Dr Bobby Stewart, a prominent soil scientist with strong contributions to dry land farming, visited IITA in 1986, and since then I worked very closely with him. Dr Stewart, founding Editor-In-Chief of the series Advances in Soil Science, invited me to be a co-editor with him. It was a great honour and privilege for me to work with him. He has been, and will continue to be, a role model for me and future generations of soil scientists.

I had strong collaboration with the International Institute of Cooperation in Agriculture (IICA) since 2020. I attended many of its events involving ministers of agriculture and high-level officials, and presented webinars on its behalf to professional organizations and the private sector (e.g. Syngenta, Pepsico). Some world-class professionals of IICA that I had the honour of working with included Dr Manuel Otero, Dr Lloyd Day, Dr Jorge Werthein, Dr Muhammad Ibrahim and Dr Francisco Mello, amongst others. It was a mutually rewarding collaboration.

In India, I had strong collaboration with ICAR (Dr M.S. Swaminathan, Dr Raj Paroda, Dr Himanshu Pathak), IARI (Dr Ch Srinivasa Rao), PAU (Vice-Chancellor Gosal), Amity University (The Founding Director Mr Chauhan, M.L. Jat and Prof. N. Kaushik) and many other agricultural universities. It is important to mention that I had the honour and privilege to host about 65 visiting scholars from institutions throughout India, and they were great professional colleagues to work with.

It is pertinent to thank the staff of the former Department of Agronomy, including Karla Gutheil, Brenda Swank, Jennifer Donovan and Laura Hugh Conover, and field staff Kenneth Scaife, Yogi Raut, Basant Rimal, Dr Raj Shrestha, Dr K. Lorenz and others. My special thanks are due to the staff and students of CFAES Rattan Lal Center for Carbon Management and Sequestration (Lal Carbon Center) and SENR for their support. Specifically, thanks are due to Maggie Willis, Hannah Shively, Alexander Jones, Kyle Sklenka, Lesley Cline and others who helped me during the preparation of this book. Figure 1.1 was drawn, photographs organized, and a few photos were taken by Alex Jones. Alex Jones also searched references for quotations that I identified and included at the beginning of each chapter.

I greatly appreciate the support from David Hemmings, Lucy Pritchard and Helen Elliott of CABI for getting the manuscript edited and for effectively managing the flow of the manuscript from me to them, and their support in the timely publication of the book.

Whatever I learnt in my life is due to the love and affection of my family: grandmother, father, uncle, Bhua Ji (aunt), elder sister Sita, cousin Mr and Mrs C.P. Sharma and brother Bhagirath Lal. The family has been a source of strength and inspiration for me.

I must convey sincere thanks to members of my immediate family including my wife, Sukhvarsha Sharma Lal, who supported me wholeheartedly throughout my career beginning in December 1971. She was always a source of strength for me and gave up her family in Karnal to travel with me to Ibadan in Nigeria and later to Columbus, Ohio. I must also thank our four children (Priya, Pratibha, Abhishek and Vivek), and five grandchildren (Juhi, Bela, Ambika, Anika and Jay) for being a source of pride and joy.

Comments

President Michael Drake

'Many years ago, Rattan Lal introduced us to the vast and complex role of soil in cultivating a healthy world. His groundbreaking research has enhanced the world's knowledge of this incredible planetary resource – one that we tend to overlook. Lal's contributions have shed light on the myriad ways soil nurtures life on Earth. His work has also uncovered soil's amazing capacity to sequester carbon and thus enhance our planet's climate resilience. As a world-renowned soil expert, and as a colleague and mentor, he has been a guide to so many on the magic and majesty of soil. I am so pleased that he put to words his many experiences and insights in this memoir, and I know his story will inspire and enlighten generations to come.'

Michael V. Drake, MD, President, University of California

Pete Smith

'Prof Rattan Lal came from humble beginnings in pre-partition India but has become the most prolific and in my opinion the greatest living soil scientist. His work has always been guided by farming realities, which were instilled in him as the child of a farming family. His long career has been studded with major global awards, including the Glinka World Soil Prize (2018), the World Food Prize (2020) and the Gulbenkian Prize for Humanity (2024). He has managed these remarkable achievements while remaining humble, supportive and willing to share his wisdom. When I was a young post-doc in the 1990s, Rattan provided me, and others like me, with opportunities to contribute to the many volumes he edited. I have learned a great deal from Rattan – as a scientist and as a person. It is an honour to know him.'

Pete Smith, Professor of Soils & Global Change, University of Aberdeen, UK

Alex McBratney

'Those of us raised in farming communities, enriched by diverse cultural backgrounds, recognize agriculture not only as a means of producing essential food for humanity but also as a way of life that shapes these communities. From such communities, great scientists sometimes emerge, driven by a deep understanding of the balance between food production and sustainability for both the environment and the people. Rattan Lal is one of these rare individuals. Rising from a small rural community in India, he has become a global leader in soil and agricultural sustainability. Rattan Lal's work spans across the globe, with significant contributions in India, Australia, Africa and the Americas. He has developed a comprehensive message and framework for soil conservation that will ensure the security of our soils for generations to come. He was one of the first to connect the vital link between soil, food security, and climate change. Through his tireless advocacy, he has brought much-needed attention to soil and soil science, raising awareness among communities and governments worldwide. His dedication to this cause is unmatched, and his achievements in publishing and communication are truly remarkable.'

Alex McBratney, Professor, Department of Soil Science, University of Sydney, Australia

Bal Ram Singh

'I met Rattan first time at the Indian Agricultural Research Institute (IARI) in 1964, where we both studied for our MSc degree and I have fond memories of staying in the same hostel and having our meals in the vegetarian cafeteria of IARI. After completion of our degrees, we both moved further abroad for our professional careers in our respective fields and left India for good. However, we have been keeping in contact with each other through our meetings in various congresses, conferences and annual meetings of ASA/SSSA. From the 1990s, our scientific co-operation developed very closely through exchange of students and his project participation in my projects in Tanzania. In fact, I spent two of my sabbatical leaves in 2010 and 2020 with him and his staff at the CFAES Rattan Lal Center for Carbon Management and Sequestration, Ohio State University, and came to know more about his scientific achievements. During these visits, I also came to know his wife Sukhvarsha, daughters Priya and Boby, and sons Abhishek and Vivek, and have very fond memories of having dinner parties with them on various occasions. Since the 1990s, we have travelled many times to Tanzania for supervising and following up the projects in various places in Tanzania, and organizing international conferences at the Sokoine University of Agriculture at Morogoro. Our co-operation since the 1990s and till today has resulted in many scientific publications, both research papers and edited books. I am also glad to mention here that by recognizing his scientific contributions and achievements in developing sustainable soil management systems and educating agronomists globally to ensure food and climate security, and strengthening co-operation with the Norwegian University of Life Science (NMBU), Rattan was awarded the highest award of Honorary Degree in 2005. I need also to mention that his passion for this soil-centric philosophy has positively impacted hundreds of millions of agricultural producers and promoted and made the foundation of a peaceful world with healthy soils and healthy people. It is also worth mentioning that he was recognized as one of the top scientists in his field internationally. He was ranked no. 1, globally, among agricultural scientists by research.com in 2023. His career, spanning six decades and in six continents, recognizes the importance of intersectional co-operation among many spheres: academic, agricultural, private industry and government.

'I am glad to note that Rattan is bringing his autobiography for the benefit of not only his family and friends but also for the world's scientific community, general public and the next generation by demonstrating how a small boy from a subsistence farm in rural India turned into a soil scientist of the highest global rank, by hard work and dedication to scientific knowledge. I wish him good luck with his autobiography, good health and a long life.'

Bal Ram Singh, Professor Emeritus and Past Chair, Division 3 of IUSS, Norwegian University of Life Sciences, Faculty of Environmental Sciences and Natural Resource Management, Ås, Norway

Takashi Kosaki

'I have known Prof. Rattan Lal since 1982 when we were working together at the International Institute of Tropical Agriculture (IITA), Ibadan, Nigeria. He started his research at IITA in 1969, spent 18 years there, and finally developed an efficient crop production system with judicious use of soil resources with minimum disturbance. This was truly a paradigm shift in soil management and was a revolutionary transformation in food production with sustaining environment. This idea was transferred into research on soil organic carbon (SOC) sequestration after he moved to the Ohio State University in 1987. This made him the world's foremost soil scientist since his originality included much of the key components in carbon management for the adaptation and mitigation of climate change.

'He is a strong supporter of the concept that humanity and the promotion of brotherly relations can only be realized by sustaining soil health so that it can provide the basic necessity of all inhabitants of planet Earth. His belief that "health of soil, plants, animals, people, environment and the planet is one and indivisible" is globally recognized. He also believes that desperateness caused by perpetual hunger is the cause of civil strife and unrest throughout the world. Therefore, he is promoting the concept that fanaticism and extremism around the world can be addressed by restoring degraded soils. I am particularly fond of his belief that "the ferocious fire in the pit of an empty stomach can only be quenched by divine powers in a loaf of bread made from grains grown on a healthy soil".

'Through his research over five decades and spread over six continents, Prof. Lal has promoted world peace through ushering in a soil-centric Green Revolution. It is this concept that earned him the 35th Japan Prize on 8 April, 2019. I never forget that he was praised at the presentation ceremony with a storm of applause by all attendees including myself and Their Majesties the Emperor and Empress of Japan. His autobiography shall recall to my mind the exciting and touching moment of the ceremony as well as his history and achievement, which shall also surely be shared with all readers in the globe.'
Takashi Kosaki, Professor, Tokyo Metropolitan University, Japan

Joachim Von Braun
'Rattan Lal's story is one of an inspiring and visionary scientist whose work has transformed our understanding of soil and its critical role in sustaining life on Earth. Through science-based evidence, he has brought global attention to the immense value of soils – the fragile and life-giving skin of our planet. Lal emphasizes that soil degradation is not just an environmental issue but also a fundamental threat to food security, social justice and human rights. He has dedicated his career to identifying solutions that not only prevent soil degradation but also restore and regenerate degraded soils, ensuring a healthier planet for future generations. Though deeply connected to his roots in India, Rattan Lal is a true global citizen. His work spans continents, impacting farmers, scientists, and policy-makers worldwide. *Making of a Soil Scientist* beautifully illustrates how one individual's vision, passion and dedication can drive meaningful change on a global scale, demonstrating that science, when applied with purpose, has the power to transform lives and safeguard our shared future.'
Joachim von Braun, President of the Pontifical Academy of Sciences, The Vatican Professor Emeritus for economic and technological change, Bonn University, Germany

Karl-Heinz Fegger
'It is an honour and a great pleasure for me to contribute a few brief thoughts on Rattan Lal's autobiography. His impressive scientific life's work is not only reflected in his countless publications and lectures; his numerous networks also deserve special mention. These are networks in many countries around the world, which Rattan initiated and filled with life over several decades and which he linked together to form a global network linking people and cultures. The focus was always on the soil as the basis of life, a complex natural body and essential environmental medium, but always in a broader scientific, social and political-ethical context. His way of thinking, coupled with his unique charisma, have always fascinated me and have certainly influenced me in one way or another!

'I met Rattan Lal in person for the first time in December 2012 at the opening ceremony of the newly founded, Dresden-based United Nations University Institute for Integrated Management of Material Fluxes and of Resources (UNU-FLORES). His keynote speech at the time, "Climate-strategic agriculture and the water-soil-waste nexus", formed an essential conceptual basis for the institute. In the years that followed, I was fortunate to work closely with him on the International Advisory Board of UNU-FLORES. Under his chairmanship, FLORES gained great international visibility, to which several Dresden Nexus conferences contributed. This was followed by the award of an honorary doctorate from TUD – Dresden University of Technology in 2015. With Rattan Lal, my university, TUD, honoured a universal scientist and recognized a humanist who is still closely associated

with TUD to this day (e.g. in the Centre for International Postgraduate Studies of Environmental Management – CIPSEM). The laudatory speech by Klaus Töpfer, Under-Secretary-General of the UN, General Director of the UN office in Nairobi and Executive Director of UNEP, is unforgettable. In addition, Rattan Lal left many other traces in Germany: he was instrumental in the founding and conceptual development of the Institute for Advanced Sustainability Studies (IASS) in Potsdam. The institute, founded by Klaus Töpfer in 2009, organized the very successful Global Soil Week in Berlin in the following years. The German Soil Science Society (DBG) honoured Rattan Lal in 2020 by awarding him the "Corresponding Membership".'
Karl-Heinz Feger, Dr. rer. nat., Professor Emeritus, TUD, Dresden, Germany; Dean of Faculty, 2013–2019; President, German Soil Science Society, 2020–2023

Alfred Hartemink
'Professor Rattan Lal grew up on a small farm in India, and by hard work, a strong education, perseverance, dedication, and perhaps a bit of luck he became one of soil science's most decorated and productive members. This book tells his story from the ground up – through studies in India and Australia, and work in Nigeria and Ohio, and eventually across the globe. I know numerous soil scientists who were born and raised on a farm and that made discoveries and international careers, but I know no one who has covered such a breadth and had the productivity and the impact of Professor Lal; his legacy can be hardly overstated. Much of his motivation comes from his upbringing and a deep need to do good – good for the globe, good for its people, and the soil on which our livelihood depends. Professor Lal has also been spiritually driven to provide scientific solutions to increase crop yields, sustain our soils and climate, and make a better living for the poor. He has done that for over half a century. If all our soils were as productive as Professor Lal, we would have no problem in feeding the world.
Alfred Hartemink, Professor of Soil Science, University of Wisconsin – Madison, Department of Soil Science

1 Starting from the Ground Up

'Change your opinions, keep to your principles; change your leaves, keep intact your roots.'
Victor Hugo from 'Thoughts', found in Postscriptum de ma vie ('Postscript of My Life')

Abstract

With the partition of India in 1947, Rattan Lal's family migrated from their native home in west Punjab to east Punjab in India. This transition was a great challenge for the family whose culture and traditions were somewhat related to those of the region which was home to the once thriving but now extinct Indus Valley civilization. Because his mother had passed away when he was about one year old, his grandmother played a major role in his upbringing and in teaching him the wisdom of ancient Vedic scriptures. Climate change and soil degradation, which were among the major causes of the demise of the ancient Indus Valley civilization, were also the foundation of Rattan Lal's professional career. His choice of studying soil science was related to being brought up on a small family farm in a rural community faced with challenges of food insecurity, drought, heatwaves and the hardships of manual farming operations. His professional interests were related to what he learnt and experienced from his family and the rural community. He devoted his life to addressing the problems of sustainable soil management of all farmers with special focus on small landholder farmers, because he was one of them.

1.1 Roots among the Ruins of the Indus Valley Civilization

> 'In this moment he was just a boy – brilliant, blessed with too much power, burdened by eternity.'
> — Leigh Bardugo, *Ruin and Rising* (2015)

Rattan Lal's family is native to west Punjab. The name Punjab is of Persian origin, with its two parts (پنج, *panj*, 'five' and آب, *āb*, 'water') being cognates of the Sanskrit words पञ्च, *pañca*, 'five' and अप्, *áp*, 'water', of the same meaning (Gandhi, 2013). The word *pañjāb* is thus calque of Indo-Aryan 'pañca-áp' and means 'The Land of Five Waters', referring to the rivers Jhelum, Chenab, Ravi, Sutlej and Beas. This region was an integral part of the Indus Valley civilization. With its rich alluvial soils and diverse landscapes and eco-regions, Punjab attracted Indo-Aryans from the Caucus region who gifted humanity with the ancient language Sanskrit (mother of all Indo-European languages); Alexander the Great, who established close links between the two ancient nations of Yunan and Bharat; the founders of Islam from south-western Arabic regions with their unique philosophy and culture; and Moguls from Uzbekistan in the north-western region, who created the great architecture of the Taj Mahal and the Red Fort among others. It is this blessed land which also gave birth to the founder of Sikhism, Guru Nanak Dev Ji. This melting pot of diverse ancient cultures was also home to poets such as Waris Shaw (author of 'Heer Ranjah'). Another famous folk tale, although less famous in Punjab, was that of 'Sassi Punnu', which was written by a Sindhi Sufi poet (Shah Abdul Latif Bhittai) in the 18th century. Other popular writers of this region included Amrita Preetam (Pinjar) and Nanak Singh (Paviter Pappi). This blessed land was also proud of its great freedom fighters such as Bhagat Singh, Lala Laj Pat Rai and Udham Singh.

Agriculture has been the primary profession of Punjab since time immemorial. *A Textbook of Punjab Agriculture* was written in 1947 by William Roberts and SBS Kartar Singh. This was the region whose hard-working farmers and brilliant agricultural scientists ushered in

DOI: 10.1079/9781836990154.0001

the Green Revolution in the 1960s and made India not only food self-sufficient but also an important exporter of food grains. A major contributor to the Green Revolution, based on his pioneering work on wheat breeding during the 1940s and 1950s, was Choudhary Ram Dhan Singh. He was the last principal of the Punjab Agriculture College at Lyallpur in 1947, when the country was partitioned and Lyallpur (now Faisalabad) became part of Pakistan.

The southern part of this region, with Hindi as the dominant language, became the modern state of Haryana in 1966. It is in this place that Rattan Lal's family settled after partition of the country. This is also the region where the great epic of Mahabharata or *Kurukshetra War occurred, and mother Sita or Siyya* (heroine of the epic *Ramayana*) was absorbed by Mother Earth at the end of her earthly life. Thus, Rattan Lal and his family are a synthesis of diverse cultures, philosophies, ancient heritage and agricultural traditions.

1.1.1 Origin of the name 'Rattan'

'A gem cannot be polished without friction, nor a man perfected without trials.'
– Chinese proverb

Rattan Lal belonged to a Brahmin family, of Vasishta Gotra. The seven major Brahmin Gotras, names of the saints whose lineages included Shandilya, Upreti, Jamadagni, Gautama, Atri, Vasishta and Kashyapa. His family's caste was Bhakri from his father's side and Lakhanpal from his mother's side. His sister Sita was married in a Bhardwaj Brahmin family. Rattan Lal was the son of Pt. Jagan Nath and Smt. Krishna Wanti. He was the grandson of Pt. Mohan Lal and Smt. Utami Devi. He had an elder sister (Sita Rani) and brother (Bhargirath Lal). He was married to Smt. Sukhvarsha Sharma in 1971 and they had four children: Priya and Pratibha (daughters), and Abhishek and Vivek (sons).They have five grandchildren (Juhi, Bela, Ambika and Anika, granddaughters, and Jay, grandson).

1.1.2 *Birth Place of Rattan Lal*

Rattan Lal was born towards the end of World War II in the region of India that is now a part of Pakistan. The exact date of birth is not known because it was recorded in his *janam Patri* (birth certificate prepared by a priest) according to the Hindu calendar. However, the *janam patri* was lost during the turmoil of the partition. His native town of Karyal Kalan is in district Gujranwala of the western Punjab, and Karyal is in close proximity to the historic city of Lahore. The language of the region is Punjabi with a strong influence of Urdu and Farsi. Most educated people of the region took pride in speaking some Farsi, especially the couplets of ancient poets (Mirza Galib, Waris Shaw, etc.) and philosophers of the region (Omar Khayyam). Rattan Lal's native village was at the northern fringes of the historic Indus Valley civilization (Fig. 1.1). Until the partition of India in 1947, Hindus, Moslems and Sikhs lived in peace and harmony with one another and respected and celebrated festivals of all religions. To some, the partition of a nation based on religious beliefs made no sense. Regardless of the historic amity among diverse cultures, what happened in 1947 affected the life of millions who were made to leave their native land. Two books which narrated the stories depicting the tragedy of partition were the subject of a book entitled *Ek Chadar Maili Si* by Rajendar Singh Bedi, and *Pinjar* by Amrita Pritam.

1.2 Extinction of the Indus Civilization

'Although it was one of the first civilizations to develop a system of writing, their texts have not been deciphered so we know little about the structure and organization of their society.'
— Hourly History (2019) *Indus Valley Civilization: A History from Beginning to End*

The Indus Valley civilization created cities 1000 years ago. The Indus civilization (also called the Harrapan civilization) was a Bronze Age civilization in the north-western region of the Indian subcontinent (Fig. 1.1; Basham, 1954). It is the region where Rattan Lal's ancestors came from. This civilization existed from 3300 BCE to 1300 BCE, and matured between 2600 BCE and 900 BCE. It is known for its urban planning, buildings and streets of burnt bricks, ceramics, water tanks, and manufacturing of cotton and woolen textiles. It extended from the modern-day north-east Afghanistan to Pakistan and north-western India (i.e. Haryana, Punjab, Rajasthan). It also had trades with the civilizations of its time in the Middle East. The civilization probably ended *c.*1300 BCE because of climate change.

The family moved to eastern Punjab and was settled in the village of Rajound (dist. Karnal) in 1948, after partition of the country (Collins and Lapierre, 1975). It is in this rural community that Rattan Lal graduated from high school in 1959. He studied at Ludhiana for his BSc (Agric) degree, graduating in 1963, and for his MSc (Soils) at the Indian Agricultural Research Institute (IARI) in New Delhi, graduating in 1965.

Annual rainfall of the region decreased from 700 mm to 300 mm or less over a period of a few centuries (Fig. 1.1). Thus, agriculture based on grain crops may have become unsustainable. It may have been the ruins of this civilization that Rattan Lal often wondered about while walking over the pebble-covered mounds from his village home in Rajound to his family farm. Growing up among ruins of the Indus Valley civilization motivated Rattan Lal to understand and research these processes of anthropogenic climate change, drought, heatwave and soil degradation. Rattan Lal has been studying these issues of soil and environmental degradation ever since his first professional job in sub-Saharan Africa (SSA) in December 1969.

Fig. 1.1. Location of major cities of the ancient Indus Valley civilization (blue), which flourished between 3300 BCE and 1300 BCE. The map also shows relevant locations to Rattan Lal (red) including the location of the village of Karyal Virkan (dist. Gujranwala) where Rattan Lal was born; this village is now in Pakistan. Rattan Lal received his BSc from Ludhiana and his MSc from Delhi. The ancient city of Bukhara is located on the Silk Road. (Map drawn by Mr Alexander M. Jones).

1.3 Leaving the Ancestral Home during Partition of the Country in 1947

> 'From inside the train we heard voices travelling up and down the platform, saying, "Hindu paani", and, from the other side, "Muslim paani". Apart from land and population, even the water had now been divided.'
> — Aanchal Malhotra (2017) *Remnants of a Separation: A History of the Partition through Material Memory*
> 'This Bloody Line'
> — Sir Cyril Radcliffe, in an unusually candid moment of reflection on the borders drawn

The family left their native ancestral home in village Karyal, dist. Gujranwala, in August 1947 during the historic partition of the country when two states of India, Punjab and Bengal, were divided into two each, one half of each remaining with India (east Punjab and west Bengal) (Collins and Lappierre, 1975). Thus, Rattan Lal's family fled their home in a hurry, leaving behind all they had, and walking away to Lahore, which was at the border of the two independent nations, India and Pakistan. Being about 5 years old, Rattan Lal has only faint memories of the traumatic event except what he remembers from the stories told by his paternal grandmother (bebe) and other members of the family (his elder sister Sita Devi and elder brother Bhagirath Lal). Rattan Lal 's mother passed away when he was about 1 year old and he was raised by his bebe (paternal grandmother), elder sister Sita and his aunt, or bua ji, who was the elder sister of his father. In reality, he was brought up by three mothers (his bebe, sister Sita and his bua ji), and he was very attached to all of them.

There are two stories worth mentioning from this tragic event. The day the family left Karyal it was raining. August is the monsoon season. Rain was a good thing but it created its own challenges of walking through muddy paths. Eventually, the family arrived at a school in a nearby town (Hafizabad), where other Hindu families were also gathered. This is where Rattan Lal fell on a slippery floor and hurt his face; he was bleeding from nose. There were no medical facilities and the family had a sick child to carry under very difficult conditions. It was his uncle (taya), his father's elder brother, and other members of the family who carried Rattan Lal on their shoulders through the muddy paths and farmlands across the villages of west Punjab. This is the time when the family was leaving its ancestral home for ever to an unknown and a challenging new world.

After several days of walking, often during the night, the family finally arrived at a refugee camp set up on the outskirts of Lahore. The camp was guarded by a Gurkha regiment. It was hard to keep track of the time, and the family probably stayed there for several weeks or months. One of the stories that Rattan Lal's bebe narrated many times was about the food that the family ate after several days of not having much to eat. Bebe had seven relatively large (about 1' or more in diameter) gold rings in each ear. She gave one of these rings to Rattan Lal's father (phapa) to sell it and buy food for the family. On a lucky day, Rattan Lal's father was able to sell an ear ring and bought some wheat flour, onions and mango pickles etc. But there were neither any utensils to make a flour dough nor a cooking place to bake rotti (naan). Bebe had to improvise however she could under the circumstances and kneaded the wheat flour on a piece of cloth to make dough. Other family members made a metal baking pan (tava) from a tin canister to bake bread on a stove made from stones or bricks. The fuel was any wood or biomass that could be found. The family ate rotti with raw onions or mango pickles. However, there were no rotti left for bebe to eat. After some time, all the family members who ate rotti complained about its rather bitter taste. While everyone slept on the floor, bebe was worried about the bitter taste of the rotti and could not sleep. She was worried that, because she did not eat any, she may be the only one left alive. She had heard the rumours (probably false) about deliberate contamination of food and water. Thus, she spent all night checking on each member, especially Rattan Lal, checking that they were alive and breathing. She was relieved when everyone woke up the following morning. About 25 years after, during the 1970s, when Rattan Lal and his young family lived in Ibadan, he remembered the story about the wheat flour that had been stored under humid and hot conditions.

The family spent a few months in each of the two refugee camps, Lahore on the Pakistan side and Amritsar on the Indian side, before they were settled in district Karnal (now in Haryana,

India; Fig. 1.1). Most of the refugees who came to India from the Gujranwala districts of west Punjab were settled around Karnal, which is about 200 miles (320 km) south-east of Amritsar and about 100 miles (160 km) north-west of Delhi. About 1 year later, in September 1948, the family eventually arrived in their new home and was settled in the rural community of Rajound, in district Karnal. Rajound is about 40 miles (65 km) south-west of Karnal, 17 miles (27 km) south-east of Kaithal and eight miles (12 km) west of Assandh (Fig. 1.1).

1.4 Childhood in a Remote Village Community of Rajound during the 1950s

> 'Who are you? A simple son of the soil, as you pretend to yourself? Oh, no. You, too, are among the infirm – you are the dreamer, the madman in a madder world...'
>
> — John Williams (1965) *Stoner*

From ages 5–15, Rattan Lal lived in a one-room mud home allotted to the family in Rajound (Tehsil Kaithal, district Karnal) in what is now in the Haryana state of India. The home was door-less with no windows, a short courtyard, a flat mud roof, and mud-made stairs leading to the roof. The family slept on the roof during summer. Rattan Lal often walked from his village home to the family farm, and the footpath passed through some raised grounds or mounds covered with pebbles of fire-baked bricks and broken pieces of clay pottery. He was curious about the source of these artifacts and asked his father, uncle and elder brother, who were farmers, about the ruins. He was told that they were the remains of some ancient civilization, which vanished long ago, probably because of some natural disasters such as drought.

Rajound, a casbah/kasbah (Arabic word *qasaba*, meaning fortress or citadel), of approx. 10,000 people in 1947/48 (population of approx. 60,000 in 2024), is about 500 km south-east of Karyal Kalan and 200 km north-west of New Delhi (Fig. 1.1). During that time, this remote community was located at the junction of three dirt roads: to Kaithal in the north-west (25 km), kasbah of Poondry in the north-east (15 km) and to kasbah of Assandh (12 km) and the town of Karnal (65 km) in the east. During the 1950s, Assandh had a large mound in the middle of the kasbah with remains of homes and brick structures, which may also have been the ruins of the Indus civilization. This region has a semi-arid climate with 40–60 cm annual rainfall and deep soils of alluvial origin which are prone to secondary salinization/alkalinization, and wind erosion during summer (see Chapter 9).

During Rattan Lal's childhood, the groundwater level in Rajound was 100–150 hands ('hath' is a traditional unit of length equal to that from the elbow to the tip of the middle finger) or 50–70m deep. Manual digging of a well for domestic water use was a challenging task and there was a thick layer (20–50 cm) of concretion or caliche ($CaCO_3$ or $MgCO_3$ formed as secondary carbonates over millennia) formed through fluctuations in the water table over a long period. This layer was hard to break or cut through to reach the water table.

At the time, Rajound had neither paved roads nor running water, no electricity connections, nor an established market. Fortunately, however, it was one of few villages in the region that had a primary school or 'madrasa', and this is where Rattan Lal studied from 1949 to 1959. For doing homework during the evenings, a kerosene lamp was the only source of light. Thus, waking up early (4 or 5 am) and studying in the pre-dawn light was the best alternative. The habit of waking up early (and going to the office at 5 am) served Rattan Lal well during his later career. When at Ohio State University, Rattan Lal came to the office at 5 am every workday, and at 7 am on weekends and holidays.

During his childhood, Rattan Lal was very much attached to his bebe. In 1949, Rattan Lal's aunt or bua ji (his father's elder sister), who was settled in Ludhiana after partition of Punjab, came to visit them. Her husband had also died during the political turmoil, and she lived with her only son (Rattan Lal's elder cousin Chander Prakash Sharma) and her father-in-law. After partition, she did not know the whereabouts of her family (mother, two brothers, two nephews and one niece). In an effort to locate her family, she contacted a cloth vender who travelled from Ludhiana to southern districts of Punjab, sold clothes and travelled among these villages on a bicycle. She asked him to help locate her family who may be settled in one of these villages of southern Punjab. She had given him a description of the family (there were no photographs) and names of her two brothers and three children. Naturally, she was disappointed whenever he came back without any information on their whereabouts; but after about a year, returning from one of his trips, the vendor informed Rattan Lal's bua ji that he had finally located her family in the village of Rajound in district Karnal. Thus, Rattan Lal's bua ji came from Ludhiana to meet her family in Rajound and took Rattan (then about 5 years old) with her to Ludhiana. She got Rattan Lal admitted to a primary school near to her home. For admission to the school, Rattan Lal's date of birth was a guess by the teacher. When asked about his age, his bua ji mentioned to the teacher that he was three or 4 years old at the time of the independence and partition of the country. Thus, the year of his birth was determined as 1944, and 5 September was probably the date of admission into the school. While there was no birth certificate, Rattan Lal's janam patri (testimonial of birth of a baby boy written in Hindi by a priest, which also provided a forecast of his future life based on astrological criteria) was prepared by his grandfather or bapoo ji (Pt. Mohan Lal) who was a Hindu priest (misher, mishra or a purohit). But the janam patri, similar to other family belongings, was lost when the family fled from their native village which is now in west Punjab, Pakistan (Fig. 1.1). Thus, Rattan Lal did not know the exact date or year of his birth.

Rattan Lal completed the first 2 years of primary school in Ludhiana. In around 1951, his father came to meet his sister in Ludhiana and informed her that he would take Rattan back to the village. He also assured his sister that there was a school in the village and Rattan Lal would be enrolled in it. Rattan gained admission to the village primary school (*pathshalay* in Hindi, *madrasa* in Urdu) in 1951. It was a two-room brick structure, surrounded by a fence of thorned branches of acacia (*Vachellia nilotica* or gum arabic tree). Most of Rattan Lal's classes in the summer were held under the shade of a peopal/peepul (*Ficus religiosa*), also called a bodhi tree (the tree under which Lord Budha received enlightenment, knowledge and wisdom (bodhi), and under the sun and open sky during winter and spring. The school also contained a small run-off pond (called a duggie) and a flower-cum-vegetable garden. With the help of a gardener-cum-caretaker, students were involved in growing flowers and vegetables in the garden with irrigation from the run-off water in the duggie. The water was lifted manually with buckets and poured into a channel connected to small garden plots. Working in the school garden turned out to be an important experience for Rattan Lal in formulating his professional career in agriculture and making him a devout gardener. Gardening became his hobby and passion throughout his life (see Chapter 8).

1.5 Cultural Foundation of Rattan Lal: Stories from Vedic Scriptures Narrated by His Grandmother

> 'Satyameva jayate' ('Truth alone triumphs')
> — *Mundaka Upanishad* 3.1.6

> 'Ekam sat vipra bahudha vadanti' ('The truth is one, the wise call it by many names')
> — *Rigveda* 1.164.46

Lal's elder sister, Sita, was married in 1952 and she lived with her husband's family, who was also settled in Rajound. For about 5 years (when he was 8–13 years old), Rattan was looked after by his grandmother (bebe), whose name was Uttami Devi, and at the age of *c.*70, she did whatever she could to cook food and make sure that rotti was available when he returned from school at about 1 pm. Bebe often joined him at school, and she walked back with him to home. When back at home, she would put some water in a flat container for him to soak his feet so that they would cool down. She would then give him one or two naan breads and some homemade ghee (purified butter). He would eat the bread with a small mango fruit (native small fibrous fruit which was sucked for its juice), mango pickle, raw onion, or a dip made from sour milk or buttermilk and salt and pepper. After the lunch, bebe would narrate some stories from *Ramayana*, *Mahabharta* and the *Upanishad*, and about the family's long journey from Karyal to Rajound, while also comforting him by creating breeze from a handheld fan made from the leaves of a date palm (*Phoenix dactylifera* L.). Bebe would often tell him that his grandfather (her late husband Pundit Mohan Lal) was a Hindu priest (purohit) and he had prepared Rattan Lal's janam patri (forecast about a baby's future), and had prophesied that he will grow up to be a *vidyawan* (educated one) and maybe his successor as a priest. She would also explain to him some mantras as narrated by his grandfather, and explained to him the importance of these mantras in daily life. One of these was the Gayatri mantra, which he narrated all his life during anxious moments because it always had rather a calming effect on him whenever he was faced with a challenging situation (see Chapter 22).

1.5.1 The *Ramayana*

आहुः सत्यं हि परमं धर्मं धर्मविदो जनाः
'The wise always believe that truth is the highest form of righteousness.'

The *Ramayana* is an ancient Hindu epic composed in Sanskrit between the seventh and fifth centuries BCE (Rajagopalachari, 1957). The *Ramayana* epic is known throughout South and South-east Asia. Stories about *Ramayana* and other scriptures were narrated to Rattan Lal by his dear bebe. Among the favourite stories from *Ramayana* that he learned from his bebe was about the birth of Sita or Siyya (wife of Lord Rama and heroine of the *Ramayana*), and she was born when her father (King Janak) was ploughing the field. The king's pundits had advised that drought could only be broken if he ploughed a field on an auspicious day determined by the pundits. While the king was ploughing the field and pundits were chanting mantras from the scriptures, it started to rain. However, while it was raining, the plough got stuck against an obstacle in the soil. Orderlies of the king started digging the soil to remove the obstacle but found an urn with a lid on it. When priests removed the urn, while chanting mantras, they found that it contained a baby girl. This was indeed a miracle that exhibited that Mother Earth (soil or *terrra matre*) had given birth to a baby girl. Thus, the earth's daughter was named by the pundits Siyya, which in Sanskrit means furrow. This mythological story became the basis of Rattan Lal's keen interest in understanding the impact of ploughing on soil. Indeed, the story was part of his childhood teachings by his bebe. When the daughter of King Janak grew up, she was married to Lord Rama, as a part of *swayamvar* (choice of husband based on some act of bravery by him). Prince Rama, from the neighbouring state of Ayodhya, as the *Ramayana* narrates, lifted the special bow and hit a target which no one else could hit. Thus, Rama won the hand of Siyya in marriage because of his bravery and skills as a warrior.

Rather than being crowned as a successor to his father, Prince Rama, along with his wife and brother, Laxman, obeyed his father's wishes to go into exile and live in a forest for 14 years. King Dashrath of Ayoudhya, father of Prince Rama, was forced by one of his four wives, Kakeye, to send Rama, the elder son (from another wife named Kaushalya), into exile so that her son named Bharat (half-brother of Prince Rama) would be the successor to the throne. Prince Rama obeyed these orders and went into exile along with his wife and younger brother. It was during the exile that Sita was kidnapped by the demon king of Sri Lanka, Ravana. With the help of an army of monkeys and their commander, Lord Hanuman, Prince Rama invaded the kingdom of Lanka, defeated Ravana and liberated Sita. Lord Rama went to Ravana as he laid mortally wounded on the battlefield and sat down towards his feet with folded hands, sought his advice and requested him to share his wisdom and immense knowledge. According to the epic, Ravana counselled Rama by stating that 'things that are bad for you seduce you easily; you run towards them impatiently, but things that are good for you fail to attract you; you shun them creatively, finding powerful excuses to justify your procrastination'. Ravana went on to state that 'Kal karna so aaj kar, aaj karna so abb, pal men pralay ho gai to phir kare ga kabb' ('Whatever is to be done tomorrow, do it today; whatever is to be done today, do it now; if there is a catastrophic event, then when will you do it?'). Rattan Lal's bebe used to state the same teaching in another phrase: 'Kull naam kaal ka, jis mein awe raat' ('Tomorrow is a name for something unknown that is separated from today by a night'). With this advice, Rama, along with Sita and Laxman, returned to Auodhya and reclaimed his kingdom. It is stated in *Ramayana* that Lord Rama returned to Auodhya (now a city in Uttar Pradesh closer to the Himalayas) in a *pushap viman* (some type of a flying-machine). Interestingly, the name of the national airline of Bangladesh is Viman. The day of the arrival of Lord Rama back to Auodhya is celebrated in India and elsewhere as the Festival of Lights or Diwali.

1.5.2 The *Mahabharata*

> 'Life is a journey, and every experience is a lesson. Learn from it, and you will grow.'

Rattan Lal also learnt about the *Mahabharata* from his bebe. The *Mahabharata* is an ancient Hindu epic (Ragagopalachari, 1958), and it supposedly happened in the very region in which Rattan Lal grew up. Kurukshetra is about 50 km north-east from Rajound. It was the focal point of the historic battle called Mahabharata or the global war (world war of ancient times). During the Kurukshetra war, Lord Krishna advised the warrior Arjuna about his duties towards his people. The teaching of Lord Krishna in the battlefield is the holy book of Hinduism, *Gita*. *Mahabharata* was likely composed between the 3rd century BCE and 3rd century CE. Some parts of the epic may go back to 400 BCE or about 2400 years ago. As well as advising of his duties to his people, Lord Krishna advised the warrior Arjuna about his duties as a father towards his family, and as a citizen towards his community. This dialogue is the basis of the *Gita*.

The epic of *Mahabharata* began with the intentional burning of a palace. The palace was made of lac, and Pandwas were residing in it. It was set on fire by their cousins, Kauravas, so that the Pandwas could be killed. However, the Pandwa brothers and their wife, Drupadi, made a dramatic escape and survived. According to the epic, all five Pandwa brothers had one wife, Drupadi (a custom of polyandry was followed among the so-called Indo-Aryans who worshipped fire (or *havan*) and used the symbol of a swastika).

An interesting part of this story was the defeat of Arjuna in a game of chess with his cousins (Kauravas) and the loss of his kingdom and wife Drupadi in a gamble. Dushasan, a cousin who defeated Arjun in the game of chess, ordered that Drupadi be insulted in public and her sari be removed. At that time, the Pandwa brothers were helpless and could not protect her honour. Lord Krishna answered Drupadi's prayers and made her sari to be of infinite length, which could never be removed.

During the great world war of Mahabharata, a large number of soldiers were supposed to have died. It is also believed by the faithful that the red colour of the soil of Kurukshetra was caused by the blood of the soldiers who died in the Mahabharata. Upon seeing the death and destruction, Arjuna refused to fight, and Lord Krishna reminded him of his obligations as a warrior, and his moral duties to fight for his rights and seek justice for his people. These are the teachings of the holy *Gita*.

Another version of the epic states that the so-called Mahabharata may never have happened, and it was simply a mythological teaching by saints (rishis or mahatmas) of that era. In this context, some philosophers believed that five Pandwas refer to five senses (eyes, ears, nose, mouth or tongue, and skin). Similarly, 100 Kauravas refer to 100 temptations (vices or vikaras) which attack the human senses and disrupt humanity from its righteous path. According to *Mahabharata*, the Pandwas had a victory over the Kauravas because of the guidance received from Lord Krishna who was the chariot driver of the warrior Arjuna. If Pandwas refer to the five senses, and Kauravas to the vices or temptations, then Krishna, according to some philosophers, must be the inner voice, soul or guiding light. Arjuna won the battle of Mahabharata because he listened to Krishna. Therefore, whoever obeys the teachings of Krishna will always be victorious. Both stories from the *Ramayana* and the *Gita*, which Rattan Lal learned from his bebe, had a long-lasting impact on the young boy. During his later career as a soil scientist, he had a professional interest in studying the colour (hue and chroma) of lateritic soils of the tropics.

The fact that Lord Krishna was a cowboy (gopal) had also been a source of strength and motivation for young Rattan Lal who had the task of looking after the family cows during the summer holidays between ages of 8 and 14. After all, if Lord Krishna himself could look after cattle, why couldn't a mere mortal do the same?

1.5.3 Some fictional stories which had a profound impact on Rattan Lal

Bebe would also narrate some fictional stories such as the reason why Krishna had blue coloured skin. It was probably because Krishna, as a baby, had to fight a legendary cobra who lived in a deep well. The cobra spat his venom on Krishna, and thus, his skin turned blue. Another story was about a curse from a rishi (saint) to the Moon. When the rishi was away from home doing meditation in the Himalayas, the Moon found out that the saint's wife was alone at home. The Moon disguised himself as human, descended to Earth, and flirted with the wife of the rishi. Rishi got angry and cursed the Moon for his bad behaviour and said that the Moon will have a permanent black spot on his face, and the spot can never be erased. The rishi also cursed his wife and caused her to be transformed into a rock. When she asked for forgiveness, the rishi said that she will be forgiven and transformed from rock into a woman many centuries from now when her soul will be released from the rock and she will regain her woman 'avatar'. This transformation from a rock to a woman will happen when Lord Rama will touch the rock with his feet. The saint explained that Lord Rama will be passing through these jungles during his search for Sita. As you may recall from the *Ramayana*, Sita or Siyya was kidnapped by the demon king of Sri Lanka named Ravana, and Lord Rama was passing through these forests in search of her. When Lord Rama sat down on the rock, it was transformed into a woman again.

When Krishna was a baby, there was a severe rainstorm that caused flooding everywhere, and there was no protection for the people of Mathura from the devastating floods or *pralaya* (प्रलय signifies a period of cosmic dissolution or destruction, often associated with a great flood or deluge, representing the end of a cycle of creation and the return of the universe to a state of potential before a new cycle begins. As the story goes, young Krishna lifted a mountain with his little finger so that all people could take shelter under it and be saved from the raving floods or *pralaya*. Another popular saying that Rattan Lal learnt during his childhood was, 'Indeed, with determination and strong will power, nothing is impossible'.

The birth of Sita happened during the drought, and one of the childhood miracles of *Mahabharata* (*Gita*) was the flood (*pralaya*, प्रलय) from which the people had to be rescued.

Bebe also narrated to Rattan the cultural beliefs regarding the occurrence of solar and lunar eclipses caused by their being swallowed by demons (Rahu and Ketu), and why milk and food must be protected during the eclipses by adding leaves of dub grass (*Cynodon dactylon*, commonly known as Bermuda grass. This grass is different than the Vetiver, which was promoted by the late King of Thailand, H.M. King Bhumibol Adulyadeg. The King's birthday (5 December) has been celebrated as the United Nations World Soil Day since 2013; see Chapter 16.)

A favourite story for Rattan Lal was that of a boy named Sarwan Kumar. Sarwan was devoted to his blind parents and carried them on his shoulders by using a carrying pole (also called a shoulder pole or a milkmaid's yoke). It is a yoke of wood or bamboo, used by people to carry a load. Thus, using the shoulder pole, Sarvan carried his parents from one religious shrine or pilgrimage to another. However, Sarwan was accidently killed by King Dashrath, father of Lord Rama, who was shot by an arrow while taking water from a pond for his blind parents. In agony, and while refusing to drink water from King Dashrath, the blind parents cursed Dashrath to suffer a similar death when his son (Lord Rama) would go into exile in forest for 14 years. Complete devotion to parents was a strong moral of this story.

Deep in his heart, Rattan Lal had always been thankful to his bebe for teaching him the cultural and spiritual values of Hinduism based on ancient scriptures. Bebe would often narrate these stories to him with accompanying *bhajans* (prayer songs), especially at bedtime, as sort of lullabies. She also held the kerosene oil lamp in her hand so that Rattan Lal could study and do his homework.

1.6 Bebe's Departure from the Earthly World

'The soul is neither born, nor does it ever die; nor having once existed, does it ever cease to be. The soul is without birth, eternal, immortal,

> and ageless. It is not destroyed when the body is destroyed.'
> — *Bhagavad Gita* 2.20

Bebe would often pray to Lord Krishna and Lord Rama for her elder grandson, Rattan's elder brother Bhagirath Lal, to be able to get married so that the family would have a semblance of a decent life. Alas, her prayers were not answered during her lifetime. The elder grandson was married in 1959, about 1 year after bebe's death.

One day, in 1958, the palm/leaf fan fell from bebe's hand, and she left the Earth for her heavenly abode. Her cremation was the biggest shock for the 14-year-old Rattan, who was inconsolable at the death of his dear bebe. He was very angry when bebe's body was cremated by his father while a priest was reciting some mantras. He asked the priest, 'What in the world did he think he was doing?', and the priest advised his sister to take him away for the cremation ceremony.

After a week or more since bebe's death, Rattan Lal's bua ji came from Ludhiana to mourn the death of her mother. She decided to take Rattan Lal with her to Ludhiana. While in Ludhiana, Rattan Lal also attended the marriage of his elder cousin (Chandar Prakash Sharma) in 1958. Rattan Lal returned to the village to complete his final year of studies at the high school. This was a hard time for Rattan Lal without bebe to look after him. Since bebe was no more, and the family consisted of four males – two boys (Rattan Lal and Bhagirath Lal) and two men (father and uncle). Taya (uncle), elder brother of Rattan Lal's father, started cooking whatever he could and Rattan Lal's sister (Sita) would come occasionally to make sure that there was lunch available when he came back from school. His sister was very kind to him and helped him throughout his youth. Breakfast was either buffalo milk while father was milking, or homemade yoghurt and buttermilk along with the previous evening's leftover rotti or naan. It is important to know that, unlike cow's milk, which must be heated (pasteurized) because of the risk of tuberculosis, buffalo milk can be consumed without prior treatment because buffaloes are immune to tuberculosis. Buffalo milk is rich and has a high fat content.

In 1959, after the matriculation examination (April/May), Rattan Lal's bua ji came to Rajound and took him to Ludhiana with her. This was the year Rattan Lal left his village for ever, returning occasionally during the holidays. Thus, the family left behind consisted of three males, looking after themselves. While Rattan Lal Joined the college (in autumn 1959), 'brathery' (a community consisting of families of the same caste and those who also migrated from Pakistan) helped the family and arranged the marriage of Rattan's elder brother to a girl from another Brahmin family who had also migrated from Pakistan. The marriage was in December 1959 and Rattan Lal came home from Ludhiana to attend the ceremony. Rattan Lal was very happy that his bebe's prayers were finally answered. Yet, Rattan Lal wished that the marriage of his brother had happened when bebe was alive. The next year, when Rattan Lal came home during the winter holidays, in December, he was pleased to find that he had a baby niece. Eventually, Rattan Lal's brother had two sons and four daughters. Rattan Lal's sister had been extremely caring, kind and supportive of him. She looked after him very well during his childhood. She had two sons and one daughter, and Rattan Lal played with her elder son (Ashok) who was the baby of the family. Because Rattan Lal's brother-in-law was a high-school graduate and had a job as a village accountant, he encouraged all of his three children to complete high school and study in college in the nearby city of Karnal. Indeed, all of Rattan Lal's sister's children did very well and eventually migrated to city of Chandigarh where they had a modern home, good jobs, and even cars. In contrast, none of his brother's six children completed high-school education. They stayed in the same village, were dependent on the farm (that Rattan Lal bought and gifted to his brother) that they rented to someone else, and continued the life of poverty and illiteracy in the house for which Rattan had provided the funds for its construction in 1971. Indeed, education matters in eliminating poverty and improving standards of living. Education and empowering of girls matter strongly in transforming the society and controlling the global population. In rural communities of India, during the 1950s and 1960s, education of girls was not a high priority. In fact, Rattan Lal was the only child of the family who graduated from high school. Neither his elder brother nor his elder sister could read nor write.

1.7 Growing Up on a Family Farm: Challenges of Traditional Farming

'When tillage begins, other arts follow. The farmers, therefore, are the founders of human civilization.'
— Daniel Webster (1840) *On the Agriculture of England*

Similar to all small landholder farmers throughout South Asia, the area of Rattan Lal's family farm ranged, from year to year, between 4 and 10 acres (1.5–4 hectares). The farm was powered by one or two pairs of bullocks or male buffaloes. A part of the farm, the low-lying or depressional areas, was irrigated by a canal that carried water from river tributaries which were part of the Indo-Gangetic river systems. Canals were often dry during the summer (May–July) and there was a scarcity of water. Assigned the task of looking after cattle during the summer holidays, Rattan experienced first-hand the impacts of drought and high temperatures (often 45°C in the shade). Walking barefoot over sandy paths during summer could lead to blisters on the feet, along with being pricked by thorns from acacia (*Acacia drepanolobium*) or whistling thorn and other ziziphus (*Ziziphus mauritiana*). During the years when monsoons were delayed or failed, Rattan Lal also participated in *hawan* (fire worship), a ritual ceremony performed by the village priest as prayers to Lord Indra, the god of rain. Alas, prayers were not always answered. These dry seasons were also characterized by severe dust storms, and dust devils were a common phenomenon in the countryside. Rattan Lal also remembers the infestation by the desert locust during summer which devoured anything green in its path and especially on the ground where it spent a night and laid its eggs in the soil. eggs hatched in about 1 week, and the swarms of locust devoured whatever crop was left. Therefore, the drought-proofing of agriculture remained a research priority for Rattan Lal throughout his professional career.

Traditional ploughing by a wooden 'ard' (plough), pulled by a pair of oxen, was the reality throughout Punjab and Haryana, and elsewhere in South and South-east Asia during the 1950s and early 1960s when he was at his family farm. High compaction of soil with low-activity clays and low soil organic matter (SOM) content, trampling by hooves of cattle and the impact of torrential monsoon rains, along with widely practised communal and uncontrolled grazing, necessitated ploughing of the dense soil surface to prepare the seedbed for the summer (kharif crops of sorghum, millet, cotton and rice) and the following winter crops (rabi season for wheat, chickpea, canola and root vegetables such as carrot, daikon radish, turnip and green peas). It was the hardship and drudgery of ploughing prior to the onset of monsoons, when the air temperatures under shade may exceed 45°C, that ploughman (called *hali*) were known to become *fakir* (*faqeer* or *faqir*). *Fakir* is a term used in Sufi Islam for those who accept nomadic life and an utter dependence on Allah because of the desperateness of the situation posed by the tradition of summer ploughing. Ploughing of flooded rice paddies by bullock- or buffalo-pulled harrow (puddling) during the hot summer was also a usual experience for farmers, the ploughman walking in the mud and holding the plough, draft animals pulling the harrow to create anaerobic conditions in soil, whose structure was deliberately being destroyed by puddling. Along with his father and brother, Rattan Lal also transplanted rice seedlings in a flooded paddy field. When it got too hot while transplanting, taking a dip in the mud would bring some relief, at least temporarily. As Rattan Lal witnessed and experienced these traditional farming operations during his youth, he had always questioned the rationale and need for these practices of multiple ploughing and puddling along with the hardships associated with them (see Chapter 9). Rather than puddling, and growing rice under flooded conditions, growing aerobic rice (similar to that of wheat) is an environmentally friendly option, even if it produces somewhat lower yield than that by puddled and transplanted paddy rice.

1.7.1 Seasonal food shortages during the 1950s

'The past is never dead. It's not even past.'
— William Faulkner (1987) *Requiem for a Nun*

During Rattan Lal's schooldays, from 1952 to 1959, he has vivid memories of seasonal food shortages. It was very common that the wheat harvest of April (Baisakh month), which is the staple cereal in north-western India (to make chapati, puri, parautha and naan), would end by November/December. At this time, coarse cereals and millets (maize, sorghum, pearl millet or bajra) harvested in September/October, would be available, for making bread and other preparations. Although rice (both basmati and dhan) was also harvested at this time, it was not a food staple for the bread-consuming population of Punjab, Haryana, Rajasthan, Uttar Pradesh and adjacent states of north-western India. During the winter when wheat was in short supply and corn cobs (unhusked) were brought home, dry cobs with grains were boiled and eaten for dinner with some salt, lemon or even ghee. In this regard, Rattan Lal was also a 'spoilt child' because he would refuse to eat bread made from maize and khichri (rice and mung beans cooked together) made from pearl millet or sorghum. Relatively high nutritional quality of millets, known to the rural community, were at that time neither understood nor appreciated by Rattan Lal. During his later career, Lal learned that millets (forgotten or orphan crops of sorghum, pearl millet, finger millet, along with pulses of urd, mung beans, chickpeas both desi and kabuli) were indeed a better diet than wheat or rice. Being ignorant and biased, he had a preference only for bread made from wheat and that, too, from the special wheat variety (C-591) developed by the famous wheat breeder of that time, Rao Bahadur Chaudhary Ram Dhan Singh. Coincidentally, this was also time of the year when everyone else in the village had a short supply of wheat. Rattan Lal's father was concerned about the attitude of the 10–12 year-old difficult child, and he would often go to the shops and community members to buy wheat for Rs 5 (around 10 kg at the time), which would suffice to meet Rattan Lal's need of naan bread for a month or more. Of the many families contacted, some would agree to sell wheat from their seed stock so that the spoilt child could eat naan made from his favourite wheat, C-591. Every household also had a hand-operated stone grinder, and Rattan Lal's sister Sita would grind a small amount of wheat when needed to make naan from the flour just for Rattan. Other members of the family (elder brother and sister, father, uncle and grandmother) would eat bread made from maize with sagh of sarson (inflorescence of canola before it bloomed), and also khichri made from pearl millet. When the Sonora wheat varieties, Mexican brown wheat bred by Dr Norman Borlaug and Dr M.S. Swaminathan, were introduced in India during early 1960s, the general population had a similar response about its low acceptance because of the preference for naan made from C-591 wheat. Thus, farmers grew about one acre of low-yielding C-591 for home consumption and the remaining land was planted to Sonora varieties for sale in the market. Similar to C-591 wheat, farmers would also grow about one acre of basmati rice for home consumption and the remaining land was cultivated to high-yielding rice varieties introduced in the Green Revolution package for sale in the market.

1.7.2 Some known and unknown, and seen and unseen, threats and fears

> 'What are the dangers of being lost in the wilderness forest?'
> – *Peddling Magazine*

> 'Use all your senses – sight, hearing, smell, and even touch – to detect dangers before they become immediate threats.'
> – *Declining God*

One of the common practices to maintain soil fertility was to have cattle rest at night in a part of the field where the dung and urine would presumably improve soil fertility. The cattle were then moved on to another patch of land. This practice could also be combined with cattle from neighbouring farms. Thus, among the family tasks assigned to Rattan Lal during the summer included taking the evening food for father, uncle and brother, who would guard the cattle, and doing other farm operations. The footpath from the village to the farm, about a half to one mile, passed through a field of millet and sorghum. Depending on the rainfall amount and distribution during the monsoons (July–September), these crops were often 6–8 feet (1.8–2.5 m) tall, especially at the flowering stage. Passing through the fields of sorghum or millet, which may be 300–400m long, was a test of the courage and mental strength of the ten-year-old. Carrying a stick, made from acacia or neem tree, was often a source of courage and strength. The threat of the unknown was mostly from snakes rather than any wild animals. Snakes also preferred the shade of crops or trees. Reciting the Gayatri mantra, while passing through those fields was important to keep his mind off the danger.

Staying for the night on the farm during the monsoon season was quite an experience. This was also the mosquito season, and getting bitten by mosquitoes caused malaria. Whereas traditional medicine was drinking some concoction from neem leaves (which was bitter), the medicine from the local dispensary was quinine. There was not much difference between the bitterness of neem and that of quinine.

Rattan Lal often took his cattle to a pond for water, and the colour of the water usually changed from muddy yellow during the monsoons to green with algae growth as the season changed into autumn (September–November), winter (December–February), spring (March–May) and late summer (June–July) prior to the rains and during the harvest season of summer (kharif) crops. Farmers often filled their water containers (made from a dried-out gourd) from the same pond from which the cattle drank, with the difference that they often filtered the water through a piece of cloth. This precautionary step of filtering out the algae and other floating matter was followed widely, but even the filtered water had the greenish colour from dissolved substances (including some microbes that were invisible). Rattan Lal preferred to get his drinking water from a bore hole dug in the dried-out canal bed. Within one or 2 hours of digging a 50 cm or deeper bore hole in the sand, some water would seep into the hole. This water was fresh, clean and tasty. Sadly, however, it was not enough for his cattle.

There were also some other invisible threats hidden around the water pond and under the surrounding shrubbery and grass. These were the cobras, hidden under the foliage, presumably in search of frogs, rats and other small rodents. While taking his cattle to water, Rattan Lal often encountered cobras. Having a stick was not enough to minimize the danger for a barefooted youth. When encountering such a danger, he often fled to where his father, uncle and brother were working about a quarter of mile away. After being assured by his brother and father that cobras do not stick around at one place, Rattan Lal would reluctantly return to his cattle, always carrying a stick in his hand.

Drought and heat, aggravated by dust storms, were known and visible threats. These were anticipated and expected as part of the lifestyle in harsh environments. The hottest months of the year were May to July, just prior to the rains in late August through September. Some of these problems were aggravated by the lack of shoes or the much-needed head cover. Acacia thorns were not the only problems; high temperatures of the sand and the bare soil surface were multipliers of the hardship for anyone, but especially for an 8–12 year-old. Dust storms were common prior to the monsoon rains, and huddling under a shrub or a tree was the only shelter available. Presence of the thorns, especially of the acacia and bakhra (*Combretum indicum*) or the Rangoon or Burma creeper, were the known threats to get used to. Coming home limping was a signal to his sister or someone else to get the thorn out of his foot. It was often removed by a safety pin or a small sowing needle.

1.7.3 Pleasant memories of the field tasks

> 'My happy rice field memories include community singing; eating "maluto"...'
> — Dann Pantoja (2014) *Childhood Memories in the Rice Fields of Santa Maria*

> '... happy days of harvesting green corn and eating it with their family beside the plot as a reward for their hard work'
> — Flor Rivera López (2019) *Cooking Stories with Native Maize*

There were also several rewards of the fieldwork. When the irrigation canal was running, jumping in it during the hot summer was fun. Another treat was availability of wild fruits and berries. During summer, the available berries included peelu, which were from a tree called vann. Rattan Lal has never seen this tree nor its fruits since leaving the village in 1959. The other delicacies of the summer were fruits of the thorny shrub called peenju and of the neem (yellow-coloured fruits). In addition, there were also wild cucurbits (including wild melons and cucumber). Roasted corn cobs on the fire, with leaves on, was a real treat. Of course, melons were also planted in the cotton field and were often a delicious treat, including those taken from a neighbour's field.

Mixed cropping of cotton and melons, and cotton and okra were the common practices.

There were also delicacies available during the winter. Among wild berries were ber (*Zizyphus*), which were delicious. Rattan Lal often saw ber bushes in Israel and throughout the Middle East during his travels later in his career (1990s and 2000s). Among cultivated delicacies of winter months were daikon radish, turnip, carrots and sugarcane. Eating carrots cooked in sugarcane juice, while making gur (native sugar made in January/February), was a rare treat. Roasted chickpeas (both desi and kabuli) were something to look for during the months of February and March. These were the rewards worth all the risks, seen and unseen, and known and unknown, for anyone, but especially for a 10–12 year-old. Although no one in that community knew what chocolates and other delicacies were, found in the cities, no one cared less about those unseen and unavailable delicacies.

Sarson (canola) sag was also a treat of a sort. Sag, a green vegetable made from florescence (called sarson ka sag), is a popular dish among the rural and urban population, and among both rich and poor. Radish (daikon), turnips and carrots (often grown on ridges marking the boundary of the wheat field) were always fresh and a delicious treat.

1.8 Lessons from his Father

> '1. sacrifice... 2. hard work... 3. never compromise on your values... 4. respect... 5. family is everything...'
>
> — *Life Savior Thoughts: Life Lessons from a Father to His Son*

Being the youngest of the three siblings the baby of the family, Rattan Lal was also close to his father. His father, Sh. Jagan Nath, was a primary-school graduate but could read and write only in Urdu, which is written in Arabic script. Most of the community members came to his father to get their letters written and read. His father was very proud that Rattan Lal was studying in the school, and the only one of three children who could read and write. While the family spoke Punjabi, Rattan Lal could read and write in Hindi, Gurmukhi (Punjabi) and, later, English. Urdu, created during the reign of the Mogul emperor Akbar, is a mixture of Hindi and Persian, and written from right to left. Thus, there was a language and cultural barrier between Rattan Lal and his father. Rattan Lal often heard his father state that 'Urdu is a great cultural language and it is a pity that it is no longer taught in schools in independent states of Punjab, Himachal and Haryana'. Yet, Urdu was taught in Uttar Pradesh, Andhra Pradesh and other states where there was a large Islamic population.

Thus, whenever Rattan Lal received a letter from his father (while he was away from the village and was in Ludhiana, Delhi, Columbus, Sydney or Ibadan), he had to take the letters from his father to someone who could read Urdu. In Ohio, he had a class fellow from egypt who could read Urdu but did not understand it, and his accent and pronunciation of key words was different. In Ibadan, Nigeria, letters were read to Rattan by a colleague from Palestine. Similarly, when Rattan Lal wrote a letter to his family in Hindi, these were taken to someone who could read it. His father's letters were poetic and affectionate. Every letter began with the same sentence in Farsi/Urdu: 'Toole umar, noore chasham, baad umur drazi ke wajay ho ki yahan pur har terah se khariyat hae auur thumari kariyhat rabb se nek chahta hoon' ('Dear Rattan and apple of my eye, I wish you a very long life. I am pleased to state that we are all happy and healthy and pray to God for your good health and happiness.').

Rattan Lal was often asked by his teacher to make a short presentation in Hindi at the school prayers in the morning. Thus, Rattan Lal would make a presentation on topics such as hard work, self-confidence, patriotism, independence, etc. His father would often share with him some couplets in Urdu that contained some words and quotations from Farsi (Persian). Rattan Lal would use these couplets in his presentations at the school (and later at Ludhiana), and these were very much appreciated. Some examples of these couplets are:

> 'Jagna hai jag le phalak ke saya tale, ta hashar soya rahega khak ke saya tale.' ('You must wake up while you are under the cover of the sky because you will be asleep for ever and ever under the cover of the soil.') (The word khak is a Farsi word for soil.)
>
> 'Parhe Farsi veche tael, dekho eh karma(n) da khel.' ('He has studied Farsi, but he is selling oil; what a misfortune!')

> 'Himmat kare insan to kya ho nahin sakta, woh kaun sa masla hai jo hal ho nahin sakta.' ('With strong determination and hard work, there is no problem or obstacle that cannot be successfully resolved.')
>
> 'Dabb ke wah the rajj ke kha.' ('Deep and intensive ploughing of soil will produce high yield and plenty to eat.') (Here was the cultural foundation that motivated Rattan Lal to devote most of his research career in studying the impact of ploughing on productivity.)
>
> 'Jinna gur pao unna hi mittha.' ('Sweetness of a syrup or tea depends on the amount of sugar added.')
>
> 'Jaun beej ke kanak nahin waddi jandi.' ('If you sow barley, do not expect to harvest wheat.')

His most favourite story was that of the howl of a jackal followed by the response from other jackals. It was in Farsi: the first jackal says, 'Pidram Sultan Bhood', or 'My father was a king'. The others reply: 'Tura Che, Tura Che, Tura Che' ('How does that matter to you, you are still a jackal?').

Lal's father would also often murmur his favourite tunes in Punjabi: 'Pagri Sambhal Jatta' ('Look after your turban, O farmer, don't let any one touch it with disrespect'); 'Cheera wala ek aa gya purauhana, nee maye there kamm nan mukke' ('O Mom, a turbaned handsome guest has arrived, and you are too busy to extend to him the traditional welcome'). He would also murmur 'Meri russ gayyi chhanjran wali te sadhe phha the rubb russia' ('When my sweetheart wearing her silver anklets is annoyed with me, it is like God has forgotten me'). These rather sad couplets indicated his father's loneliness; they and other quotations guided Rattan Lal throughout his studies and his professional career.

His father was happy when he was riding the bullock cart back home from the day's farm work while chewing sugarcane or eating roasted green chickpea pods (called hola), eating freshly harvested carrots, daikon radish or turnips.

His father used to say that even Jawaharlal Nehru (the first Prime Minister of India) would not be so happy as a farmer was when riding his bullock cart, or taking a nap under a tree with the big clod of soil as his pillow. Riding a plank being pulled by bullocks to level an irrigated field just before ploughing or seeding was a joyful ride for farmers. Rattan Lal, when 8–10 years old, would also sit on the plank, while holding on to his father's legs so that he would not fall.

1.9 Rattan Lal's Mother and Her Family

> 'Nothing hurts as much as a void in the heart, a craving for the touch that never falls apart.'
> — Blaze Buds (2023) responding to 'What could be some inspirational quotes for orphans or children who lost one parent?'

Rattan Lal's mother's name was Krishna Wanti, and she died when he was about 1 year old. It was about the same time when his grandfather (Pt. Mohan Lal) also died. He did not have any memories of either one of them, and there were no photos of any family members. Whenever Rattan Lal asked his grandmother about his mother and what she looked like, his bebe replied that he looked like his mother and that she was very caring and gentle. Because of his mother's passing and departure from their native village of Karyal Kalan, Rattan Lal had practically no contact with his mother's family, which comprised his two uncles and three cousins. Only one of his maternal uncles was married, and he had three children (two boys and one girl, and the girl was the eldest of the three). After partition, his mother's family settled in a village name Ravar, which was about four miles south-west of Karnal. The first time he ever visited his mother's family was when he was in the ninth grade of high school. The first time that Rattan Lal met his maternal uncle was when he and the family came to attend the marriage of his sister Sita in 1955. Rattan Lal's cousin's sister (Prakash Rani) was also married in a family settled in Rajound, but Rattan Lal had minimal contact with her cousin sister because he left his native village in April/May 1959. No, or little, contact with his mother's family was partly because of strained relations between his father and his maternal uncle (his mother's elder brother). Whenever his father and maternal uncle met, they always argued about some money or cattle borrowed by the uncle from his father but not returned. The last time Rattan Lal met and saw his maternal uncle was during his own wedding in 1971 in Karnal. Rattan Lal did not get a chance to travel to the village of Ravar except once in late 1957.

Consequently, he lost contact with his mother's family altogether.

1.10 Hindu Festivals of Dussehra and Diwali

> '[Hindu] festivals promote moral values, reflect nature's cycles, and offer opportunities for personal reflection, social bonding, and spiritual growth.'
> — Rudraksha bazar (2024) *Hindu Festivals and Their Significance*

During his childhood, Rattan Lal eagerly awaited annual celebrations of the Hindu festival of Dussehra. The festival was preceded by a village drama called 'Ram Leela' or the story of Lord Rama. The roles of women were played by men, who would dress up in women's clothes (e.g. Sita, the wife of Lord Rama, Kaushallya, the mother of Lord Rama, Kakeye, the second queen or stepmother of Lord Rama, and Swaroop Nakha, the sister of the Drawid King Ravana of Lanka). In the evening, under the light of lamps and mashals (kerosene-fired torches), villagers played out the story of the childhood of Lord Rama, his marriage to Sita, the tragic decision of his father for Lord Rama to go into exile for 14 years (along with his younger brother Laxman and his wife Sita), the kidnapping of Sita by the demon king Ravana of Lanka, release of a woman from stone when Lord Rama rested on it, the war (aided by the monkey king Hanumana) to liberate Sita, and the demise of Ravana. One of the favourite stories about Ravana was that he did not die soon after he was defeated. He presumably had ten heads and a new head would sprout whenever one was cut off. Thus, Ravana could not decide when to die. Finally, he made the decision about the day when his soul would leave his body and start a new life in another body – the theory of reincarnation believed by Hindus, Sikhs, Buddhists, Jains and other religions related to the Vedic culture. For that auspicious moment, the famous saying in Hindi is, 'Kull karna so aaj kar, aaj karna so abb; pal mein pralay ho gai, to phir kare ga kabb' (Whatever is to be done tomorrow, should be done today; whatever is to be done today, should be done now. O fool, if the world comes to an end, when will you get these important things done.').

The festival of Dussehra was celebrated with the cremation of the effigies of Ravana and his son. One month after the festival was the celebration of Diwali or the Festival of Lights, when Lord Rama returned to his capital of Auodhyya (now in Uttar Pradesh) and his return was celebrated with the lighting of oil lamps in all homes and distribution of sweets among family members and friends. Rama and his entourage returned from Lanka to Auodhya (near the Himalayas) on a flying-machine called *pushap viman* and arrived in 1 day (so says the writing in the *Ramayana*).

Two other festivals celebrated in all villages were Holi, the festival of colour, and Lorhi, the celebration of the birth of a son in the family. The Lorhi festival is celebrated during the winter (January). It involves a week of festivities. In many ways it is similar to Halloween in the West. During the Lorhi week, boys and girls go to homes during the evening chanting folksongs and asking for gifts. An example of the chanting in Punjabi is as follows:

> Sunder mundriyye, Ho
> Tera kaun vichara, Ho
> Dulla Bhatti wala, Ho Dulle,dhee viahi, Ho
> Ser shakkar pai, Ho
> Shakkar kaun smete, Ho
> Akho Mundio Dherni, Ho

Similar traditional chanting was done by girls and they all enjoyed the gifts or treats received from the families. The most common treat was popcorn (baked in hot sand in a pan heated by a wood fire) and gur (traditional or homemade raw sugar) which was then eaten while sitting around a bonfire. On the final day of Lorhi, everyone went to the home of a family who had been blessed in receiving a baby boy during the previous year and sang traditional songs of congratulation. A party of men and women also went to the same house but their demands were different (e.g. bottle of local rum, for the men). The women of the household and neighbours gathered in the courtyard of the family blessed with a baby boy, sat around a bonfire and sang songs wishing for good health and long life of the baby while eating sweets and popcorn. The women were usually dressed in their finest clothes and jewellery. This was a memorable festival.

Then there was the festival of Baisakhi. It is celebrated on the first day of the month of Baisakh (around mid-April). It is a spring festival and this is the day when harvesting of wheat is started in Punjab and Haryana. The ceremonial harvesting of wheat was done on the first day of Baisakh even if the crop was not yet fully matured. Each village had a *mela* (a fair or a gathering to meet and celebrate the good harvest). There would be traditional games played by boys and men such as wrestling, kabaddi, folk dances such as bhangra by men and giddha by women, etc. There were also shops to buy and sell traditional merchandise and many freshly made sweets. The most popular sweet of the Baisakhi festival was jalebba or jalebbi. It is made from fermented wheat dough squeezed through a narrow outlet (made from a leaf or a cloth) and then deep fried in mustard (canola) oil to form concentric rings. The fried jalebbi was then dipped in super-saturated syrup of sugar. When freshly baked, this is the most delicious sweet. Rattan Lal learnt during his travels (2000s to 2020s) to the Middle Eastern countries that jalebbi is the common dish in all Arabic countries. It is the national dish of Kuwait. Other sweets made on the Baiaskhi festival included halwa (a dish similar to that in egypt and elsewhere in the Middle East), and the gulab jamun are made from dried milk. Gulab jamun turned out to be the national sweet dish of Turkey. Then there was the laddoo; it is made from chickpea flour. Laddoo is probably the only native sweet dish of north-western India; all others seem to have been adapted from the Middle East or *vice versa*. Any time Rattan Lal travelled to the Middle East, later in his career, he enjoyed eating these foods. Once, in Algiers, Rattan Lal mentioned to a woman professor that his childhood favourite sweet was jalebba. The next day she brought him a big box of freshly made jalebba and insisted that he eat as much as he can and take the remainder back to Ohio.

1.11 Sharing the Harvest for Support Services from Village Communities

'Traditional South Asian farming often involved village communities or extended families to pool resources and labor for cultivation and distribution, ensuring food security and social cohesion.'

— *Community Based Farming in South Asia*

Village life involved a strict division of labour in terms of the support services. Most households were farmers, comprised of small landholders, but there were also a few large landowners. The village where Rattan Lal's family settled after the partition of the country was dominated by the so-called Raj Poots. These were the traditional ruling class, and some households were large landholders with hundreds or thousands of acres. One of his class fellows from one such family was Badan Singh. He was fair-complexioned and tall with light brown hair and light yellow eyes. He talked about his joint family (during mid-1950s) of about 50 members. They had a large brick mansion with a huge courtyard containing hundreds of cows. The name of his father was Dhaula Singh (or 'white lion'). Similarly, there was another large landholder who had a dark complexion and was appropriately named Kala Singh ('black lion'). The names of people reflected their complexions and did not mean anything else. There was a dark-complexioned woman and her name was Kali Behan ('black sister'). Among his class fellows, there were a dozen of boys from such families and they were natives of the village (rather than migrants, as was Rattan Lal's family). The language, dress and food habits of these natives were different to those of Rattan Lal and his community of migrants from Pakistan.

The farming community, small landholders and large farmers, had support services from non-farming families. These families were those who provided services such as water from the village well (two or three pitchers of water every morning), barbers, carpenters who made traditional beds and tools for churning yoghurt (for separating ghee), pottery makers and shoemakers. At the end of the year when wheat and chickpea were harvested (April), these families received some grains (about 40 kg of wheat for each support family). These families also received some special gifts during marriage or (male) childbirth. This was a unique system. Later, when Rattan Lal was working in Africa (early 1970s) and went to visit his family, he asked his brother if those traditional services were still available. He was informed that they were breaking down, and all farmers and others pay cash for every service required (e.g. haircuts, having shoes made etc.).

1.12 Community Life

'I always said that to get India stronger, first strengthen its roots, and all our roots are our country's interiors, rural areas and villages.'
— Raveena Tandon (2011) 'All we need is the will'

'I want to turn the clock back to when people lived in small villages and took care of each other.'
— Pete Seeger (1995) 'The old left'

There was a strong co-operation among the community or the 'brathery'. For Rattan Lal, the brathery comprised Brahmin families who migrated from Pakistan and were all settled in the same village. These families came from different villages in Pakistan but were members of the same caste and shared a common language, culture and food habits. They helped one another in ploughing, sowing and harvesting of wheat or other crops, and in sharing the land for grazing the cattle. They also supported one another in arranging marriage of their children or naming ceremonies, and came together in both happy and sad times. For the marriage of a daughter, the community would cook food for the groom's family, arrange accommodation for the night, and contribute milk and other supplies needed for the guests. The brathery also contributed money (Rs 11, 21, 51 or 101, always an odd number) and these amounts were recorded and returned to the donor family when they had a marriage of a child. The brathery would also help one another in constructing their homes made from mud bricks and preparing a thatched roof or a cattleshed for winter. Being from the same caste (Brahmin) and the same historical background (refugees from Pakistan), the community was a very closely knit group. All elder members of the community were addressed by the younger generation as grandfather (bappoo) and grandmother (bebe), uncle (chacha) and aunt (chachi), brother (bhai) and sister (behan). The brathery also provided advice and suggestions for arranging marriages of children among these families and their acquaintances. Indeed, the marriages of Rattan Lal's elder sister and brother were arranged with the help of the brathery and they also had a strong hand in arranging Rattan Lal's own marriage (see Chapter 11).

It was this brathery or 'community' that was the foundation of the culture and traditions based on the Snatan Dharma (Vedic culture based on Sanskrit scriptures) and the ancient Vedic traditions. It was this community life that taught Rattan Lal the respect for teachers and elders, and caring for the family in a joint family system. In school, teachers were held in the highest respect, just as parents and elders were at home. Parents and teachers were given the status and respect of gods and goddesses. There could be no god or goddess bigger than the teachers, parents and the community elders. In the class (under a tree or in a room), all students would stand up when a teacher entered and remained standing until after the teacher sat down on his chair or advised them to sit down. A student must stand up while talking or addressing a teacher, and a student will not sit down until he/she is advised to do so. There is a saying in Hindi which is, 'Guru Eeish Donno Khare, Mayin Kis ke Laggun Paon; Balihari Guru Aapne, jis Govind Deo Milaye' ('If you meet teacher and God standing together, to whom would you convey your greetings first? Of course, it is the teacher who deserves more respect because without the teacher no one would have ever have reached God.').

Respecting a teacher is something unique to the Asian culture, and western society could certainly benefit by adopting this tradition.

1.13 Farewell to Village Life

When Rattan Lal graduated from the high school in 1959 and left his village and went to Ludhiana to live with his aunt or bua ji (father's elder sister), he did not realize that he was leaving the village and the community life for ever. He did not realize that the lessons learned from life in the village (from age 5–15) had left a strong impression on him and would become the guiding principles for the rest of his life.

Life in the village, which had none of the modern amenities, made Rattan Lal what or who he is, and what or who he aspired to become during his professional career. This village had no roads, no electricity, no running water, no established market (bazaar) nor any modern amenities. The nearest town of Kaithal (with a railway station and paved road) was 17 miles (27 km) away. During the monsoons, the mud road was not traversable and going back to college in Ludhiana meant getting to Kaithal by a horse tonga, mule cart or even walking. Roads were improved by the late 1960s and the village also received electricity by then.

Rattan Lal graduated from high school in 1959 and procured admission to the Government Agricultural College, Ludhiana (now the Punjab Agricultural University; see Chapter 2). During one of the visits to his family in Rajound, he was asked by the village elders about the specific things that he was learning in college. They were surprised when Rattan Lal replied that he was learning the science of how and why to plough, the names and properties of different types of soil, and the identification of weeds, insects and pests. The community members (brathery) told him, 'You did not have to go to the college to learn how and when to plough; we could have taught you better in this very village. We hope and pray that you do not become 'fakir' simply while learning how to plough during hot and muggy summers.'

When leaving the village in 1959, Rattan Lal did not realize that this was the beginning of a new chapter in his life. He was leaving behind a community that gave him his identity and the isentity of who he was to become. In all respects, even though he left the community, the community and its lifestyle and social values never him. The culture, social values, thinking, and way of life were embedded in every cell of his body. These social values and their customs stayed with him on his journey through six continents and over six decades. These guiding principles of his life, which served him and his profession of soil science very well, are listed below:

- vegetarian eating habits, without any alcoholic beverages, smoking or substance abuse;
- hard work and strong willpower, dedication and complete commitment to the task at hand, loyalty to the job, and doing the best that could be done under the site-specific situations;
- respect for parents, elders and teachers;
- respect for law and order;
- care and reverence for the soil as the source of all essential ecosystem services.
- respect and reverence for nature; and
- traditional knowledge about agriculture and judicious management of soil, water and natural resources (Rattan used this knowledge to his advantage during the professional career).

These principles were never to be forgotten, and will stay with Rattan until his soul leaves to enter that of another living being, and his present body becomes soil, which he studied all his life and hopes to do the same in the next life.

1.14 Motivaton and Inspiration of a Budding Soil Scientist

The historical, cultural, family and ethnic background of Rattan Lal were the foundation, inspiration and motivation of his research into conservation tillage or conservation agriculture (CA); drought and scarcity of water; soil structure and its ramifications including crusting and compaction; water run-off and soil erosion; soil degradation including salinity and sodicity; and, finally, global warming and climate change. Herein lies the foundation of his soil-centric approach and lifelong desire to enhance the prestige of agriculture and the soil science profession; to alleviate drudgery and hardship faced by all farmers, but especially those of the resource-poor small landholders; and to bring about transformation in agriculture and make it a solution to global issues of the 21st century. His strong belief that sustainable management of the world's soils is the foundation for world peace and prosperity. After all, the once-thriving civilization (Indus or Harappan), the roots of his culture and identity, vanished because of climate change and soil degradation. Rattan Lal never forgot this reality of the past history of his family and community.

Bibliography

Bardugo, L. (2015) *Ruin and Rising*. Square Fish.

Basham, A.L. (1954) *The Wonder that Was India*. Sidgewick and Jackson. ISBN: 033043909X

Bedi, R.S. (1986) *Ek Chaddar Maili Si*. Hindi Pocket Books, Gurugram, Haryana, India. ISBN: 9788193935552

Bhagavad Gita (E. Easwaran, trans.; 2nd edn) (2007) The blue mountain center of meditation.

Blaze Buds (2023) What could be some inspirational quotes for orphans or children who lost one parent? Quora. Available at: https://www.quora.com/What-could-be-some-inspirational-quotes-for-orphans-or-children-who-lost-one-parent (accessed 26 May 2025).

Collins, L. and Lapierre, D. (1975) *Freedom at Midnight*. Collins.

Easwaran, E. (2007) *The Upanishads*, 2nd edn. Nilgiri Press.

Faulkner, W. (1987) *Requiem for a Nun*. Garland Publishing, New York.

Gandhi, R. (2013) *Punjab: A History from Aurangzeb to Mountbatten*. Aleph Book Company, New Delhi. ISBN: 9789383064410

Hourly History (2019) *Indus Valley Civilization: A History from Beginning to End*. Independently published.

Hugo, V. (2020) *Post-scriptum de ma vie*. Available at: https://www.gutenberg.org/ebooks/63768 (accessed 26 May 2025).

Life Savior Thoughts (2023) Life lessons from a father to his son. Available at: https://medium.com/@lifesaviorthoughts/life-lessons-from-a-father-to-his-son-9ba48ab1e8a5 (accessed 26 May 2025).

López, F.R. (2019) CreativeVoice: Cooking stories with native maize. Terralingua. Available at: https://terralingua.org/stories/creativevoice-cooking-stories-with-native-maize/ (accessed 26 May 2025).

Madhvani, R. (2017) This Bloody Line. Available at: https://www.youtube.com/watch?v=alvQ2AyCCu4 (accessed 26 May 2025).

Malhotra, A. (2017) *Remnants of a Separation: A History of the Partition through Material Memory*. HarperCollins Publishers.

McIntosh, J. (2008) *The Ancient Indus Valley: New Perspectives*. Bloomsbury Publishing. ISBN 9781576079072

Pantoja, D. (2014) Childhood memories in the rice fields of Santa Maria. Available at: https://waves.ca/2014/04/17/childhood-memories-in-the-rice-fields-of-santa-maria/ (accessed 26 May 2025).

Pritam, A. (2009) *Pinjar: The Skeleton and other Stories*. Tara Press. ISBN: 9788183860970

Rajagopalachari, C. (1957) *The Ramayana* [Abridged Retelling]. Vedic Book Series. ISBN: 8172763654

Ragagopalachari, C. (1958) *Mahabharata*. Bharatiya Vidya Bhavan. ISBN: 9788172763688

Report of the agricultural meeting, held in Boston, January 13, 1840, containing the remarks on that occasion of the Hon. Daniel Webster, of the U.S. Senate, and of Professor Silliman, M.D., L.L.D., of Yale College, Conn.: with Notes by Henry Colman, Commissioner for the Agricultural Survey of the State (1840) Salem Power Press.

Rigveda: Shakala Samhita (1990–1997). Delhi vedic trust, New Delhi.

Roberts, W.S.B.S.K. and Singh (1947) *A Text Book of Punjab Agriculture*. Civil and Military Gazette Ltd, Lahore, Pakistan.

Rudraksha bazar (2024) Hindu festivals and their significance. Available at: http://rudrakshabazar.com/insight/detail/hindu-festivals (accessed 26 May 2025).

Seeger, P. (1995) The old left. *The New York Times* 22 January. Available at: https://www.nytimes.com/1995/01/22/magazine/sunday-january-22-1995-the-old-left.html (accessed 26 May 2025).

Singh, N. (2020) *Pavitar Papi*.

Tandon, R. (2011) All we need is the will. *The Times of India*. Available at: https://timesofindia.indiatimes.com/blogs/raveenatandonblog/all-we-need-is-the-will/ (accessed 26 May 2025).

Williams, J. (1965) *Stoner*. Penguin.

2 The Making of a Soil Scientist

'Throughout human history, our relationship with the soil has affected our ability to cultivate crops and influenced the success of civilizations. This relationship between humans, the earth, and food sources affirms soil as the foundation of agriculture.'
— Sanjai J. Parikh and Bruce R. James (2012) *Soil: The Foundation of Agriculture*

Abstract

Rattan Lal's move from the village of Rajound to Ludhiana, his admission into the college, and support for his education were entirely supported by his aunt (bua ji). Just as his grandmother supported him when he studied in the village, his bua ji was the source of strength and motivation for his college life in Ludhiana. Challenges of obtaining admission to the college also strengthened his resolve to excel and prove to himself that being the best in his village school implies that he can also be the best student in his college. The goal of obtaining Rs 20 per month as a scholarship was the strong motivational force for him to study hard and become the best student of his class. The choice of soil science as a major field of study was guided by his personal experiences of growing up on a family farm, performing farm operations such as ploughing by bullocks driven hard, learning the importance of soil from Sanskrit scriptures and directly from traditional farming, and by the mentoring of some excellent professors at the colleges in Ludhiana, Delhi, Columbus and Sydney. He was also encouraged by his professors at Ludhiana and Delhi to seek admission to the Ohio State University.

2.1 The Great Support and Guidance From Bua Ji

Just as Rattan Lal's bebe was the source of strength and joy for education in the village school, he was indeed fortunate and highly blessed in having support of his bua ji (father's elder sister) during his college days. Indeed, she was the mother and the well-wisher who looked after him and all his needs for 4 years when he stayed at Ludhiana in her home, especially for food and a comfortable home in which to live. She was kind and thoughtful and made sure that his basic needs were met. Indeed, she was responsible for his graduation from the college and for his becoming a professional in soil science.

The credit for getting a college education entirely goes to his bua ji and her family. It is also important to thank her son and Rattan Lal's elder cousin (Sh. Chander Prakash Sharma) and his wife (Santosh Sharma). Both were very kind and supportive of him. It is rather sad that Rattan Lal lost contact with both after the death of his bua ji in 1974. She was his great source of strength and inspiration.

2.2 Transitioning to a College Student

Rattan Lal obtained a matriculation degree from the Government High School in Rajound, and was the 'first-class first' in his graduating class. This was the popular term when someone obtained the highest grade and passed in the first division (60% of the score according to the old system of grading). In the handwritten letter of reference, in addition to Rattan Lal's grades and public speaking, the headmaster felt it pertinent to also state his interest in some games such as football (soccer). However, Rattan Lal had never played any games at the school, especially soccer. With this diploma and letter of reference, Rattan

DOI: 10.1079/9781836990154.0002

Lal travelled to Ludhiana, where his bua ji lived, and he stayed with his aunt's family for 4 years during his college studies from 1959 to 1963. Rattan Lal's journey from Rajound to Kaithal (17 miles), where there was a railway station, was often by horse-driven tonga, mule-pulled cart, or on foot. For the 15-year-old, these were fun rides.

Soon after arriving in Ludhiana, Rattan Lal got acquainted with a neighbour of similar age named Bhim Sen. The latter had a clerical job at the newly established Government Agricultural College and Research Institute. The College of Agriculture (hereafter the 'college') was the sister-institute of the similar college established in 1905 at Lyallpur, Faisalabad, now in west Punjab, Pakistan. Sikh and Hindu faculty, who left Pakistan after partition of India, established the college at Ludhiana. During 1957, the U.S. Agency for International Development (USAID) provided a long-term grant to the Ohio State University (OSU) to help develop this college as a Land Grant University (LGU) (Propp, 1968; Mehta *et al.*, 2017). Thus, there were OSU faculty at the college and Bhim Sen told Rattan some humorous stories about American professors from a state called Oyyo (Ohio) in the USA (i.e. driving car, changing the car tyre by themselves, drinking coffee rather than tea, and speaking English in an accent hard to understand). One day, Rattan Lal asked Bhim Sen if he could get a similar job at the college, but Bhim Sen did not show any enthusiasm about this idea, and he merely replied 'What for?' The college was 4 miles (6.5 km) from where Rattan Lal and Bhim Sen lived. Neither of them had a bicycle and there was no public transport between the city and the college.

Upon further insistence by Rattan to visit the college, Bhim Sen agreed that he may visit and see for himself how American professors come to the office in a car and teach classes. Impressed as he was with all the fairy-tales narrated by Bhim Sen, Rattan Lal remained focused on getting a clerical job similar to the one that Bhim Sen had. After all, Bhim Sen was earning Rs 60 per month (approx. $20 at that time). To Lal , who was completely broke, Rs 60 per month was a lot of money. However, his bua ji and Bhim Sen insisted that Rattan Lal should apply for admission and earn a college degree. Bhim Sen also introduced Rattan to the head clerk of the college (Bhumi Chand), who informed them that the top five students in the class got a scholarship of Rs 8 per month, and those who are certified refugees from Pakistan also received an additional Rs 12 per month. The lure of earning Rs 20 per month was a game-changing idea and Rattan Lal was now thinking about getting admission to the college and earning Rs 20 per month. Indeed, he was already daydreaming about being a student at the college. This transformation, brought about by the lure of receiving Rs 20 per month as a scholarship, was noticed by Bhim Sen. While walking back home from a visit to the college, Bhim Sen could see that the country boy was already flying high, daydreaming all the time of being a college student, and not even worried about being struck by a bicycle *riksha* nor worried about being run over by a horse-pulled tonga.

Finally, with the lure of a scholarship and persistent encouragement from his bua ji and Bhim Sen, Rattan Lal completed the application forms for admission to the college for the classes beginning in August 1959. He was thrilled to be invited for an interview. He dressed in his home-washed traditional countryside clothes (pajama and a long shirt reaching below the knees and both pressed by folding and being weighted down) and wore jutti (shoes) made by the village shoemaker from a raw cowhide. Even by that time, Rattan Lal also had a small pony tail (*chhoti*) as symbol of religious devotion of being a Brahmin. Both Rattan Lal and Bhim Sen jogged to the college and arrived in time for the interview. When his name was called, Rattan Lal went inside the interview room. Unfortunately, however, dressed as a village boy, he was not considered a potential student. He was told to leave the room and send the next candidate in. When asked by Bhim Sen how the interview went, Rattan Lal replied that the panel did not ask him any questions. Bhim Sen assured him not to worry –it was because he was the 'first-class first', so there was no need to ask any questions. Alas, Bhim Sen was wrong, because when the names of those selected for admission were announced, Rattan's name was not listed among the lucky 75 elite students selected for the class of 1963. All those dreams were shattered. Rattan had neither a job nor admission to his dream college. His despair and desperation were obvious and irreconcilable.

Both Rattan Lal's bua ji and Bhim Sen were hopeful, however, and assured him that God would help and that 'there is nothing to worry about'.

Just opposite to the college entrance, on the Gurgaon Road, was the office-cum-residence of the local congressman or member of legislative assembly (MLA) of the state. He often had town hall meetings with his constituents on specific days. At the encouragement of Bhim Sen and his bua ji, Rattan Lal went to see the MLA, showed him the high-school diploma and the handwritten letter of reference and narrated his disappointing experiences with the interview. The MLA was sympathetic and promised to arrange a meeting with the state Minister of Agriculture (Giani Kartar Singh). The visit with the minister was encouraging, and the minister promised to bring this issue to the attention of the Chief Minister of Punjab (Sardar Partap Singh Kairon). With the final approval of the Chief Minister and the Minister of Agriculture, Rattan Lal was admitted into the class of 1963 at the Government Agriculture College and Research Institute. His class roll number, by which students were known rather than by their name, was 83. In other words, Rattan Lal was the 83rd student to be admitted into the class. It is also important to know that in 1959 there were neither any female students nor any women staff in the college. While handing over the letter of approval for admission, the minister reminded Rattan Lal that if he was as good in the college as he was in the village high school, he would hope to see Rattan Lal during the annual prize distribution function of the college (Fig. 2.1). That was obviously a great challenge that Rattan never forgot.

Rattan Lal joined the classes about one month later than the other students, and jogging to and from the college (6.5 km or 4 miles) was for him among the greatest pleasures and challenges of those days. He also studied hard. To be away from the street noise and loud radios

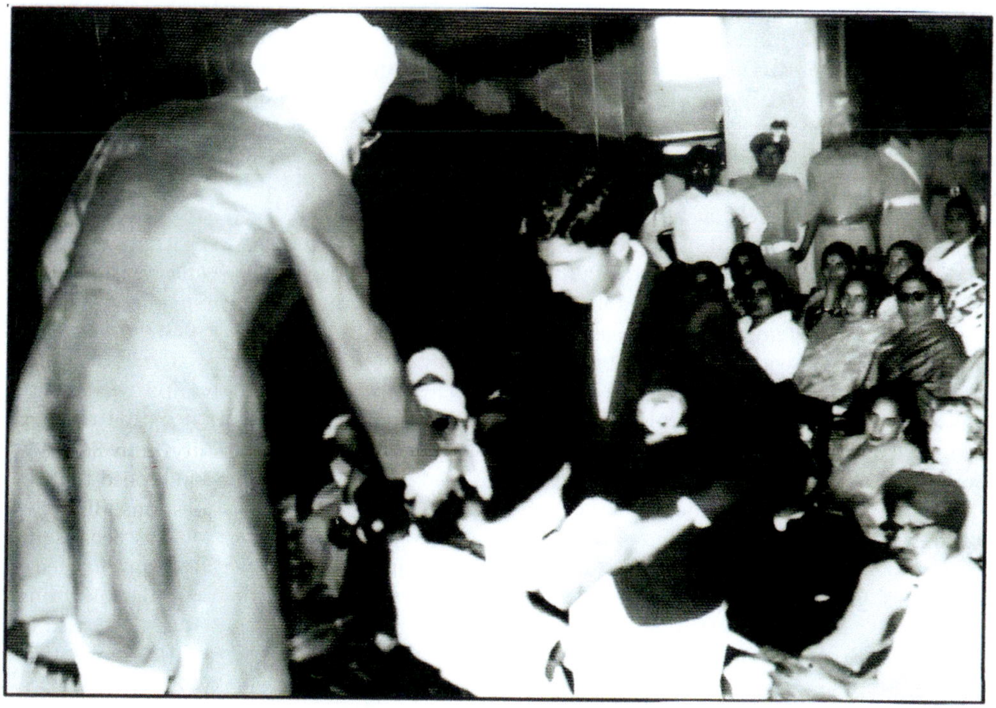

Fig. 2.1. Sh. Partap Singh Kairon, Chief Minister of Punjab, presenting an award to Rattan Lal for academic distinction in his class in 1960 at the Government Agricultural College and Research Institute, Ludhiana, Punjab, India. (Photo: PAU photo album)

played by neighbours during the day and in the evening, Rattan would often walk to a nearby park (called Committee Bagg) or go to a farm at the outskirts of the town to study. Early in the morning (4 or 5 am), especially during summer, studying under the street light was a rewarding experience, and he got his homework done just before jogging to the college.

During the first few weeks, Rattan Lal also developed friendships with his classmates including Sat Pal Magu, Kishan Chand Dhawan and Balraj Singh Parmar. Later, Kishan Chand changed his religion, and became Kishan Singh Dhawan. Another student, who was in the final year of his BSc, was Partap Kumar Khanna. Sat Pal, Partap Kumar, Balraj all joined IARI in New Delhi to pursue graduate degrees.

To his great surprise, Rattan earned the highest grade for most courses offered during the freshman year. Indeed, the minister *did* recognize him during the award ceremony (Fig. 2.1). In addition to the academic awards, Rattan Lal also received awards in the poetry competition (in Punjabi and Hindi) and in the debate contest (in English and Hindi).

This was a day of celebration and thanks to goddess Sarasvati, because Rattan Lal received a scholarship of Rs 20 per month. He was given a total of Rs 240, an amount that he had never seen nor held in his hand. Finally he had some money, and he was happy. He also purchased a second-hand bicycle for Rs 80 so that he could ride to and from the college. Finally, dedication and hard work paid dividends that he had hoped for. Above all, he gained confidence that it was possible even for a cow boy from a remote village to earn good grades and compete with the elite and affluent city boys. A good boy in the village proved to be good also in Ludhiana. It was this confidence that turned out to be even more important for his future career than the financial reward.

At debate/poetry competitions in other colleges in Ludhiana and the township of Gurgaon (about 25 km from Ludhiana), Rattan Lal won some books. One of the books was entitled *Ek Chaddar Maili Si* ('The Scarf Was Dirty'), by Rajendar Singh Bedi), and another was called *Pinjar* by Amrita Pritam. Rattan Lal read these books, which vividly explained the plight of the young girls kidnapped during partition of the country. When the Red Cross rescued them and they were reunited with their families, it was their families that refused to accept them because 'their scarf had been made dirty'. These books, along with the stories narrated by his bebe, had a profound effect on the young teenager. He became rather aloof and focused more on writing and reciting poetry than on studying agronomy, plant physiology, weed science and basic chemistry or physics. One day, his academic advisor in the college noted this change and advised Rattan to forget the poetry and focus on his studies. Writing and reciting poetry may not be good enough to get you rotti, kapra aur, makan (bread, clothes and accommodation). Rattan Lal promised his advisor that he would do as was suggested. In another debate, he received a poetry book, *Gitanjali*, by the great poet Rabinder Nath Tagore. It was a great joy to read this book.

Yet, there were numerous ups and downs in the college life. One activity, the agricultural fieldwork (i.e. agronomy, involving how to plough, do the weeding and identify predominant weeds) had a dress code of 'wearing khaki only and not any white clothes'. Rattan Lal had only one pajama, which unfortunately happened to be white. Thus, on the day of the agronomy fieldwork, Rattan Lal borrowed the striped night-pajama from his cousin. The teacher was not pleased with this outfit, and Rattan Lal got reprimanded for his rude behaviour in coming to the class wearing a night pajama. At the end of the class, Rattan Lal explained to the teacher that he had no choice but to wear whatever he could get from his cousin. Fortunately, the teacher was understanding and sympathetic.

The second defining incident occurred during Rattan Lal's very first month after joining the college. This was related to the handwritten reference letter from the school headmaster which mentioned Rattan Lal's interest in playing soccer or football. However, Rattan Lal had never played football nor had any interest in it. The college athletics group sent a letter to all those whose certificate mentioned interest in football. Rattan Lal indicated to the organizers that he was not interested in football. Nonetheless, he had to be on the field as a part of the practice game. In the very first few minutes of the practice game, someone kicked the ball and it hit Rattan directly in his stomach, and down he went, rather unceremoniously. This incident ended any remote possibility of Rattan Lal's potential career in any athletics.

The third incident was in a zoology class and it was attributed to Rattan Lal being a vegetarian (vegan diet but with yoghurt and cheese), a habit that he followed throughout his life. One of the labs of zoology involved dissection of a pigeon. Being a strict vegetarian, Rattan Lal was reluctant to do this. Following the guidelines of the instructor about fixing the wings of the bird with nails to the dissection board, Rattan Lal proceeded nervously to secure the pigeon. However, the bird was apparently not properly treated with chloroform, and started flipping its wings. Shocked and frightened by this, Rattan Lal passed out in the class, and that incidence ended any remote possibility of him pursuing a career in zoology or in medical college. So what was left to major in included soil science and agronomy.

Among the positive parts of the college life was the helpful nature of the professors. One day Rattan Lal was walking back from the college to the city on a very hot summer afternoon. On the way, Rattan was asked by his professor in botany to sit on the back seat of his bicycle and he gave him the ride to the city. This was just one example of the courtesy, care and helpful nature of the faculty.

Punjab was famous for some of its pre-Green Revolution soil scientists including Dr D.R. Bhumbla, Dr J.S. Kanwar, Dr N.S. Randhawa and Dr G.S. Sekhon among others. Most of them were appointed to high-level positions at the Indian Council of Agricultural Research (ICAR) in New Delhi and they were role models for Rattan, who was interested in studying soil science.

2.3 Majoring in Soil Science

Rattan Lal's best subject during the first 3 years in college was botany. It was taught by professor R.S. Vasudeva and Dr H.R. Kalia. The latter had returned from OSU with a PhD degree, and he was indeed a good teacher. Rattan Lal was also taught agriculture communication and extension by Prof B.S. Samundri. The last name, Samundri, referred to that of the village, near Lyallpur in west Punjab in Pakistan, where the professor's family came from. Professor and Mrs Samundri were in the first batch of the faculty who were sent to OSU in 1955 to receive a higher degree, and both graduated from OSU in 1957. Prof Samundri was fond of narrating memorable stories about the college in Lyallpur, Pakistan, and OSU in Columbus, Ohio. Prof Samundri later became the founding Vice-Chancellor (President) of the Guru Nanak Dev University in Amritsar. In addition to him planting an idea in Rattan Lal's head of studying abroad, it also was a great coincidence that India's Ambassador to the USA, from 2018 to 2022, Hon. T.S. Sandhu, was the son of Prof and Mrs Samundri.

Several alumni of OSU, who were Rattan Lal's professors at the college, made him aware of the far-away college in Oyyo (the Ohio State University) in a country called the United States of America. One of those alumni of OSU was Dr D.R. Bhumbla, a professor of soil science who had just returned from Ohio in 1961. Similarly to Prof Samundri and many other faculty, Dr Bhumbla had also started his career in Lyallpur, Pakistan, and went to OSU to study soil science. Rattan Lal was a student in the very first class in soil science that Dr Bhumbla taught at the college on his return from OSU in 1961. During the senior year, each student had to decide the subject of specialization. Rattan Lal discussed his preference for botany and plant breeding among his close friends and class fellows. However, Dr Bhumbla encouraged Rattan Lal to specialize in soil science, and he became Rattan Lal's mentor and role model.

The popular textbook for the soil science class was *The Nature and Properties of Soils* by Lyon, Buckman and Brady. There was only one copy of this rare book in the college library and there was always a long waiting list for it. When Rattan Lal finally got his turn to get the book for 2 weeks, he decided to do something about it. There were no photocopying facilities in 1960/61; thus, Rattan Lal copied the entire book in his own handwriting while sitting under a tree in a park or on a farm during the day, and under the street light during the summer nights. He had a sore arm at the end of the 2 weeks, but he was the proud owner of the textbook written in his own handwriting! About ten years later, when he joined the International Institute of Tropical Agriculture (IITA), in 1970, Dr Nyle Brady (then Director General of the International Rice Research Institute, Los Banos, Philippines) visited IITA. Rattan Lal

Fig. 2.2. Rattan Lal received the gold medal at the first ever commencement at PAU in April 1963. (Photo: PAU photo album)

asked Dr Brady if he was the co-author of the book. Dr Brady responded, 'Yes, have you read it?' Dr Brady was pleased but also confused with Rattan Lal's response: 'Yes, sir, I wrote the book! The handwritten book was deposited in the PAU library in 1963 when Rattan Lal left Ludhiana for New Delhi.

Rattan Lal completed his BSc from the college and received the degree in April 1963, in the very first class which received a diploma from the Punjab Agricultural University (PAU). It was now a university established and created along the lines of land grant universities (LGUs) of the USA, such as OSU. Rattan Lal was the valedictorian of this historic class, and his name was written on the board of the Hall of Fame for the class of 1963 (Fig. 2.2), and he was awarded the gold medal (Fig. 2.3).

2.4 Linking Soil Science to Sanskrit Scriptures

Rattan Lal's interest in soil science is also attributed to his family background, and being brought up on a subsistence farm using traditional agricultural methods. He was strongly influenced by his bebe's ideology and by listening to her stories from scriptures, especially the birth of Siyya (baby girl Sita) when King Janak was advised by pundits that ploughing of a field was needed to break the long drought that had plagued his kingdom. The problems of facing drought and heatwaves, surviving dust storms, and lack of sufficient amounts of safe water for him and his cattle were always in his thoughts. During his later career, especially as faculty

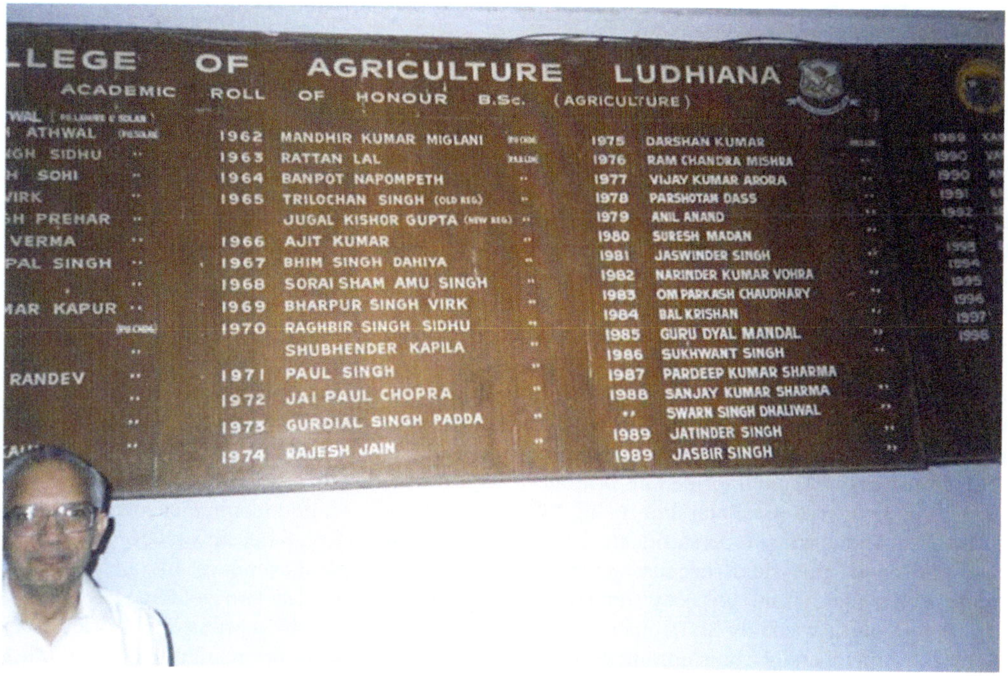

Fig. 2.3. Academic Roll of Honor for Rattan Lal for the class of 1963 at PAU, Ludhiana, India. (Photo: Dr David Hansen)

member at Oyyo (OSU) for almost 40 years, from 1987 to 2027, Rattan Lal was influenced by several quotations from Sanskrit.

Some examples of these quotations which he used in his writings and presentations are listed below:

> *Vasudhaiva Kutumbakam*, which is a Sanskrit mantra found in Hindu texts such as the *Maha Upanishad*. It means 'The World Is One Family'.
>
> कृषतिजिल पावक गगन समीरा पंच तत्व से बना शरीरा : The human body is made of five elements: soil, water, energy, space and wind. This is the basis of the 'One Health Concept'.
>
> Pavan Guru Paani Pita Mata Dharat Mahat: Prayers from Gurubani in Sikh religion, stating that wind is guru, water is father, and the greatest of all is Mother Earth.
>
> *Prakriti Pratyay pranya poratam*, or 'one life form changes into another'. While it sounds similar to Darwin's theory of evolution, this is the karma philosophy. Change of life form one type to another depends on the karma or the deeds done in the previous life.

2.5 A Rockefeller Scholar

Soon after graduating from PAU, Rattan Lal received a job offer of Rs 100 per month from a private fertilizer company. The job offer had a five-year contract to be signed. Rattan Lal sought the advice of Prof Bhumbla who suggested that he did not to sign the contract and rather apply for admission into the Postgraduate School of the Indian Agricultural Research Institute (IARI), also called the Pusa Institute, located in New Delhi. For the Soil Science and Agricultural Chemistry Division, there were five seats in the MSc class for which there were several hundred applicants. This is the second time that Rattan Lal was faced with the daunting challenge of another interview. But this time, in addition to differences in dress and shoes, he also had more confidence. The Director of the Rockefeller Foundation in India, Dr Ralph Cummings, Sr, was also on the interview panel. Dr Cummings was a PhD student of Dr Richard Bradfield at OSU during the late 1930s. Dr Bradfield established the soil physics lab, which Rattan Lal managed after joining OSU as a faculty member in 1987 until his retirement. Dr Bradfield moved from OSU to Cornell in 1937 and his graduate student Ralph Cummings Sr also transferred to Cornell with his professor. When the results of the interview at IARI were announced, Rattan Lal was one of the five students admitted into the MSc class with major in Soil Science and Agricultural Chemistry. He was also awarded the Rockefeller Foundation assistantship of Rs 200 per month. The Assistantship was exactly ten times the amount he received at Ludhiana as the best student of the class. It was hard to believe that he would receive Rs 200 per month while also pursuing his MSc degree. When he returned to Rajound and informed his brathery that he had received a scholarship of Rs 200 per month, there was, understandably, much suspicion about this claim because even the salary of an agriculture inspector (extension agent) at the time was only Rs 100 per month. Many elder members of the community asked his father whether Rattan Lal was telling the truth.

Among Rattan Lal's advisory committee members for his MSc degree were Dr T.D. Biswas, Dr R.K. Chatterjee, Dr Guy Baird from USAID, and Dr Bill C. Wright from the Rockefeller Foundation. Lal's thesis project was 'Physical Properties of Soils of India Under Corn'. During the Green Revolution era, which was also initiated during the early 1960s, the Rockefeller Foundation was focused on increasing cereal production in India, and the Cereal Lab at IARI was named the Cummings Laboratory. The study of physical properties of soils of India was the first research project that Rattan Lal ever conducted. While studying at IARI, Rattan Lal also came to know of Prof M.S. Swaminathan, a famous plant breeder, who was at that time Head of the Department of Botany. Dr Swaminathan was working with Dr Norman Borlaug (Nobel Peace Prize Laureate of 1970) to launch the Green Revolution in India. One of Prof Swaminathan's students in botany was Rajendar (Raj) Singh Paroda. Dr Raj Paroda later became the Director General of the Indian Council of Agricultural Research (ICAR) and was a good friend of Rattan Lal. Both Dr M.S. Swaminathan and Dr R.S. Paroda received honorary degrees from OSU.

While Rattan Lal was studying at IARI, Prof Bhumbla advised him to apply for admission to OSU for a PhD degree. In addition to Prof Bhumbla, Dr B.C. Wright and some professors from IARI also provided letters of reference. Rattan Lal received admission at OSU along with a Graduate Research Assistantship of US$2640 per annum from the Ohio Agricultural Research and Development Center (OARDC). When he discussed the admission at OSU with Dr Wright, the latter suggested that he could have tried getting admission for a PhD at Cornell (where Wright had received his PhD degree). Rattan Lal was surprised and responded that he did not know what the term 'Cornell' meant; he only knew about Oyyo (OSU) and had no idea what was the so-called Cornell.

The convocation (commencement) at IARI was scheduled for 28 December 1965. With the highest grade point average of his class at IARI, Rattan was awarded two gold medals at the convocation by the then Minister of Agriculture of India, Shri. C. Subramenium (Fig. 2.4). The convocation was attended by his family from the village (father, brother, sister-in-law, sister, brother-in-law and his bua ji). Some of them had never been to New Delhi before. They stayed at the house of Sh. Om Prakash Sharma who was also from Rajound (of the same Brahmin community who migrated from Pakistan) and the only one at the time from the village who was also a college graduate. Thus, Rattan Lal always addressed him as 'bhai sahib' or elder brother.

When writing his MSc thesis, Rattan Lal asked Om Prakash Sharma if he knew of someone who could help him type his thesis. Mr Sharma replied positively, and typed his thesis every evening after returning from work, and did not charge him any money. Mr Sharma remarked that 'an elder brother cannot take money from a younger brother'. During Rattan Lal 's later career at IITA, Mr Sharma was posted in Paris as a staff of India's embassy, and Rattan Lal visited him and his family in the mid-1970s in Paris.

Rattan Lal had booked a flight to Ohio with departure on the evening of the 28 December 1965. During the transit stop at the Santa Cruz airport in Bombay (now Mumbai), Rattan Lal changed Indian rupees into dollars. At the time, the Reserve Bank of India permitted changing of money at the airport to a maximum amount of US$8 (Rs 40) for foreign travel. Thus, Rattan

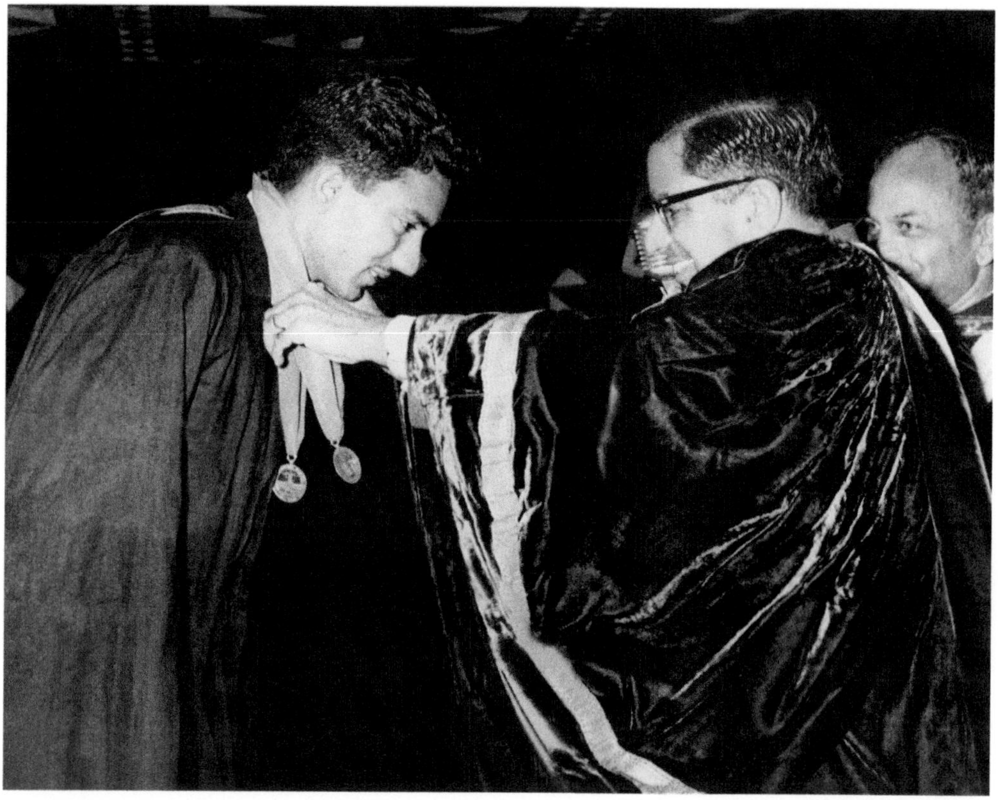

Fig. 2.4. Sh. C. Subramenium, the Minister of Agriculture of the Government of India, presenting two medals to Rattan Lal for being the best student of his class at the Indian Agricultural Research Institute, New Delhi. (Photo: Lal Family)

Lal arrived in Columbus, Ohio, on 29 December 1965 with all of US$8 in his pocket, along with a strong determination to face the challenges of the new world and to harness all the opportunities it offered.

2.6 Becoming a Buckeye

Among the new challenges in Ohio were the cold weather in December/January, poor skills in both written and spoken English, and weaknesses in basic sciences including mathematics and physics. A winter coat was the first necessity. He visited the international office on the opening day of the classes and the office staff offered him a long overcoat for US$1. It was very comfortable. Rattan Lal had to begin maths and physics classes from the very beginning, along with the study of English for foreign students. Among all courses, those in soil science were comparatively easy. Similarly to his experiences in the village school, Rattan Lal liked studying maths, and enrolled in a total of nine classes in maths, one in each trimester that he was enrolled at OSU. At one point, he discussed the possibility of majoring in maths rather than in soil physics, and his advisor (Prof George S. Taylor) persuaded him to 'stick with soil science'. His thesis research was to assess the effects of soil drainage (water table depth) on composition of soil air (carbon dioxide, methane, oxygen, nitrous oxide content and their diffusion or emission into the atmosphere), mineral nutrient uptake, and corn growth and grain yield (Figs 2.5–2.7). This experience in measuring gaseous composition of soil air and emission of these gases was extremely useful to

Fig. 2.5. The drainage and water table depth experiment established at the The Ohio State University farm in 1967 (courtesy of the field helper in agronomy who took these photographs in 1967 and gave them to Rattan Lal).

him by the time climate change became a buzz word during 1990s and later. By then, Rattan Lal was being prepared to face the challenge of understanding the 'soil-climate change nexus'.

The data showed high concentration of CO_2 and CH_4 in poorly drained soil. These observations on gaseous emissions were made in 1967, about 30 years before the anthropogenic warming became an issue.

In addition to Prof George S. Taylor, there were two other professors of soil physics at OSU: Prof L.D. Baver and Prof David M. Van Doren, Jr; the latter was based at OARDC, Wooster, Ohio. Along with Dr Glover Triplett, Dr Van Doren initiated the world's first no-till experiments on three sites in Ohio during 1960 (Wooster, Hoytville and South Charleston). The fourth site under long-term no-till was established at Coshocton by USDA-ARS. Thus, Rattan Lal had some first-hand experiences in no-till farming (now called conservation agriculture (CA). It was this experience that helped Rattan Lal start some of the first no-till experiments at IITA in West Africa during 1970. It is the data from these experiments which documented CA's effectiveness in controlling soil erosion even on steep slopes of up to 15%.

Prof Baver was the author of the first-ever textbook written on soil physics. He was also leader of the OSU team at Udaipur, India. Thus, Prof Baver was the source of inspiration for Rattan Lal to co-author a book on *Principles of Soil Physics* in 2004. Rattan Lal completed his PhD degree in August 1968 and was offered a postdoctoral fellowship at the University of Kentucky and the University of Illinois, both at US$ 11,000 per annum. However, Prof Taylor and Prof Baver had other opinions: 'What new things will you learn from another LGU in the Midwest which you have not already learned

Fig. 2.6. Corn (*Zea mays*) was used as the test crop to study its response to a constant and fluctuating water table for the 1967 growing season. (Photo: a field helper in agronomy who took these photographs and gave them to Rattan Lal)

at OSU?' Meanwhile, Prof Baver travelled to Adelaide, Australia, to attend the meeting of the International Society of Soil Sciences (ISSS), now International Union of Soil Sciences (IUSS). There he met Prof N. Collis-George, Head of the Department of Soil Science, who mentioned to Prof Baver that the University of Sydney had an opening for a senior research fellow (postdoctoral fellow) and did he know of a suitable candidate? Prof Baver told him, 'Yes, indeed, I know just the right person for this position'. Thus, Prof Baver and Prof Taylor called Rattan Lal and told him to apply for the opening at the University of Sydney. The salary at the University of Sydney for this position was less than half of that at the University of Kentucky and Illinois, but Prof Baver and Prof Taylor assured Rattan Lal that he should be thinking more about his professional development than about the difference between a one-year salary in the USA compared with the same at the University of Sydney. Thus, Rattan Lal left Ohio and arrived in Sydney in early December 1968.

For the visa to Australia, Rattan Lal took an overnight bus to New York. He attended the interview at the embassy, the third interview of his career. The visa office asked him if he was related to another Lal who played cricket. Rattan Lal had no idea who the officer was talking about. The officer jokingly told his colleague that this Lal is no good at playing cricket. The other Lal happened to be captain of the India cricket team that had just defeated Australia. Ignoring Rattan Lal's lack of knowledge about cricket, his immigrant visa to Australia was approved.

While waiting for the visa to Australia, Rattan Lal was asked sometime in November 1968 to visit the Head of the Department of Agronomy at The Ohio State University (Dr Garth Volk) in his office. Dr Volk had a visitor with him,

Fig. 2.7. Effects of water table depth (poor drainage conditions) were monitored on composition of soil air. (Photo: a field worker in the agronomy Department in 1997 who took these photos and gave them to Rattan Lal)

Dr Herbert Albrecht, who was former President of the North Dakota State University and had retired to take up a position as the Director General to launch a new Center – the International Institute of Tropical Agriculture (IITA) in Ibadan, Nigeria. These details were not known to Rattan Lal during this visit, which turned out to be an interview for a possible professional position of a soil physicist at IITA. Dr Albrecht also knew that the contract in Sydney was for 1 year with a possible extension for another year.

Prof Garth Volk's son, Bob, was a graduate student in agronomy at the same time. He brought an old and non-functional television to the lab and worked on repairing it during the lunch break for several weeks. When it was functional, Rattan Lal asked Bob, now that the TV was in working order, what he planned to do with it. Bob said that he planned to sell it to Rattan Lal for US$5. In addition to his benevolence, Bob also invited Rattan Lal to his wedding reception to which he showed up directly from doing experiments in the field. This was the first time that Rattan Lal had attended a formal reception and for which he was not prepared at all. This cultural lesson of giving was never forgotten by Rattan Lal, who, in 2021, donated US$ 1 million to OSU and donated another US$ 0.35 million in 2024 after receiving the Gulbenkian Prize for Humanity in the subject of mitigation and adaptation of the anthropogenic climate change (see Chapter 15).

2.7 Competition among Graduate Students

One of the driving forces for working hard in graduate classes was the 'curve' system of grading. If the average grade of the entire class (which was considered a C grade) was high, then obtaining an A grade in that class was a major challenge. There were always one or two students in every class who would raise the class average grade. Some of these brilliant students were also in soil science classes, and these were all round best, not just in studies. Keeping up with these brilliant students was not an easy task for someone who did not understand the language well enough to communicate easily. Among these brilliant students of that time (1965–1968) were Terry Logan, Norman Fausey and Chuck Ullery. In addition, Neil Smeck and Jerry Bigham were students a few years later. During his latter career, Terry Logan was a well-known professor of environmental soil chemistry and founding director of the Environment Science Graduate Program (ESGP) of The Ohio State University. Norman Fausey was head of the USDA Drainage Unit at Ohio State. Jerry Bigham was head of the SENR and co-director of the OARDC. Neil Smeck was Professor of Pedology and soil classification. These were highly successful class fellows and contemporaries of Rattan Lal during his graduate studies at OSU from December 1965 to August 1968.

One of Rattan Lal's class fellows, with whom he took many classes, was Horace Smith. Horace was helpful to Rattan Lal in teaching him how to drive a car. In 1968, Horace had a red and very fancy car called a Mustang. During his later career, Horace was appointed as Director of the Soil Survey Division of the USDA Soil Conservation Service or SCS (later called Natural Resource Conservation Service or NRCS). When Rattan Lal returned to Ohio in the late 1980s, he met Horace Smith by chance when he was walking through the USDA building in Washington and saw his name-plate. This surprise and chance meeting led to a long-time support for the Carbon Management and Sequestration Center (C-MASC) from USDA-NRCS (see Chapter 15)

After graduating during the fall trimester of 1968, Rattan Lal had hoped to stay in the USA but the advice of his professors was something that he could not ignore. Prof Taylor remarked to Rattan Lal at a later time about his concern. During the 1960s, job opportunities in the USA for foreign students were limited, and Prof Taylor was concerned that accepting a sub-professional job may jeopardize Rattan Lal's career. Nonetheless. Prof Taylor knew about Rattan Lal's desire to procure a job in the U.S. and told him that if he proved himself to be a good researcher, it is possible that he may come back to the U.S., and may even return to Ohio. Rattan Lal thought a lot about how to 'prove himself', and this new task of proving himself turned out to be his new challenge for the coming two decades.

2.8 High and Low Tea in Sydney

Rattan Lal's career as a soil scientist began with his arrival at the University of Sydney in December 1968. Prof Baver and Prof Taylor were right, and yes, there was indeed a difference in thinking about soil science in Australia compared with that of the mid-western USA. While staying at the International House of the University of Sydney, Rattan Lal learnt the difference between 'high tea' (dinner) and 'low tea' (late-afternoon snack). During the hot summer in December/January, he also learnt the cultural importance of wearing a neck tie, even while wearing a short with a half-sleeved shirt. There were also differences in approach to studying soil science with more emphasis on basic processes at the University of Sydney than in Ohio and the mid-western USA. Even the system of soil classification was different. Prof Collis-George was a physical chemist with a strong background in basic sciences including physical chemistry, physics and mathematics. His writing style and preparation of manuscript were superb.

The thematic focus of Rattan Lal's research in Sydney was understanding physical and thermodynamical processes which govern water infiltration into Vertisols, dominant soils of the semi-arid tropics (i.e. Australia, central India, Sudan, Mexico). Vertisols contain montmorillonitic clays, have a high swell-shrink capacity and low water infiltration rate, and are prone to accelerated run-off and soil erosion. Based on the laboratory research, Rattan found that 'heat of wetting' evolved during rapid wetting of a dry Vertisol was the cause of low water infiltration rate. Thus, keeping the soil mulched so that it does not get excessively dry may be one of the solutions to improve water infiltration rate. During his one-year stay, Rattan Lal published five journal articles on this theme with Prof

Collis-George while also learning how to write research articles in a Cambridge style, from where Prof Collis-George had come. Indeed, Prof Baver and Prof Taylor were right about suggesting Rattan Lal to take a post-doctoral position at the University of Sydney.

Rattan Lal often travelled to Narrabri, about 520 km (324 miles) west of Sydney, to conduct fieldwork on Vertisols. He worked with Bryan Bridge who was doing his PhD degree and with Coffee Laryea (from Ghana, who had done his BSc at The Ohio State University), who was doing an MSc degree with Prof Collis-George. Dr Laryea returned to his native Ghana after completing his degree and was later Chair of the Board of Trustees of ICRISAT in Hyderabad, India.

The fieldwork was a unique experience for Rattan Lal to study water infiltration rate by using a double-ring infiltrometer for 3 hours at a time. Mariotte bottle technique was used to maintain a constant water level in the inner ring. Two-dimensional infiltration test was done in Vertisols by using a circular-cavity technique, and visual advance of wetting front was observed on the soil surface over time. After completing the infiltration test, the study site was covered with a plastic sheet to minimize losses of soil water by evaporation. Soil samples from different depths were obtained for determination of moisture content at 24 or 48 hours after to determine field-moisture capacity. These data, similar to those for determining the infiltration rate in laboratory column, were used to fit the Philps equation and determine soil-water sorptivity and transmissivity.

The field visit to Narrabri had some other memorable experiences. The outdoors, including farmland, was infested with house flies. There were swarms of flies in the field, and it was useful to wear a screen mask. Rattan Lal never understood the origin of these flies, which were harmless and merely a nuisance.

A few months after studying Vertisols at Narrabri in New South Wales, Prof Collis-George invited two famous soil scientists to review Rattan Lal's research project: Dr Dennis J. Greenland (originally from the University of Reading, UK, and on sabbatical leave in Adelaide) and Dr T.J. Marshal, Chair of the Wheat Industry Research Council and author of the book *Soil Moisture*. This chance meeting with Dr Greenland turned out to be a long-term collaboration with him and the University of Reading during the later part of Rattan Lal's career.

Bibliography

Bridge, B.J., Collis-George, N. and Lal, R. (1970) The effect of wall lubricants and column confinement on the infiltration behavior of a swelling soil in the laboratory. *Australian Journal of Soil Research* 8, 259–272.

Collis-George, N. and Lal, R. (1970) Infiltration into columns of swelling soil as studied by high speed photography. *Australian Journal of Soil Research* 8, 195–207.

Collis-George, N. and Lal, R. (1971) Infiltration and structural changes as influenced by initial moisture content. *Australian Journal of Soil Research* 9, 107–116.

Collis-George, N. and Lal, R. (1972) The temperature profiles of soil columns during infiltration. *Australian Journal of Soil Research* 11, 93–105.

Lal, R. (1973) Effects of methods of seedbed preparation and time of planting of maize in western Nigeria. *Experimental Agriculture* 9, 303–313.

Lal, R., Bridge, B.J. and Collis-George, N. (1970) The effect of column diameter on the infiltration behavior of a swelling soil. *Australian Journal of Soil Research* 8, 185–193.

Mehta, S., Lal, R. and Hansen, D.O. (2017) U.S. Land Grant Universities in India: Assessing the consequences of agricultural partnership, 1952–1972. *International Journal of Agricultural Development* 53, 58–70.

Parikh, S.J. and James, B.R. (2012) Soil: The foundation of agriculture. *Nature Education Knowledge* 3, 2. Available at: http://www.nature.com/scitable/knowledge/library/soil-the-foundation-of-agriculture-84224268

Propp, K.M. (1968) The establishment of agricultural universities in India: A case study of the role of USAID-U.S. University Technical Assistance. U.S. Agency for International Development. Available at: https://pdf.usaid.gov/pdf_docs/PNRAB325.pdf (accessed 28 May 2025).

3 A Good Village and College Life: The Unique Experience that Shaped the Future

'Village life offers a unique perspective on living, characterized by simplicity, deep-rooted traditions, and a strong sense of community.'
– Nabeel Khan (2024) *Village Life: A Deep Dive into Community, Tradition, and Sustainability*

Abstract

Both village and college life provided Rattan Lal with a unique opportunity to experience challenges and face hardships. He was able to transform these hardships into opportunities to study hard and be successful in his career. These formative years were the foundation which shaped his professional life and provided him with ideas to be addressed through research in sustainable management of soil and agriculture. These were the environments which taught Rattan Lal how to face challenges with determination and willpower. In comparison with the village, some of the available amenities in the city of Ludhiana provided a unique experience for him. There was a street light, and one of these was in front of his bua ji's home. There was a municipal water tap about 200 m away, and there were libraries in the college and the city from which to borrow books and learn. These amenities indicated to Rattan Lal that the 'sky is the limit' to achieve academic accomplishments and the only limitation was himself and his ability to work hard.

3.1 The Rural Upbringing

Rattan Lal had a unique experience in his 15 years of village life (4–5 years in Pakistan and 10 years in Rajound, followed by 4 years of college at Ludhiana, 3 years in Ohio and 1 year in Sydney). He was proud of his family and community heritage. These formative years were the foundation which shaped his future career. Village life, simple and basic, was the greatest opportunity for Rattan Lal to experience the reality of growing up on a subsistence farm and living in a community which looked after everyone and took care of their collective needs. This is where Rattan Lal experienced drought, heatwaves, dust storms, low crop yields, seasonal food shortages and lack of proper medical facilities, which had a long-lasting effect on his research career and were the focus of his soil-centric approach to addressing global food and nutritional insecurity. These were the environments which taught Rattan Lal how to face challenges with determination and willpower. Achieving the highest grade point in his tenth grade at the village school, with meagre facilities and no laboratories, gave him the much-needed confidence that he could accomplish whatever he desired as long as he had total commitment towards that goal. Achieving the highest grade-point average was the driving force that also got him first position in his BSc and MSc degrees and later in his career when he was globally ranked no. 1 in soil science, agronomy and environmental sciences by research.com in 2023 and 2024, and by the Stanford Ranking in 2024.

The first 19 years of his life, in the village where he completed the high school, followed by 4 years at Ludhiana, where he completed his BSc degree, were indeed formative years which shaped his future. Since he was the top student in his village class, he believed that he could also be the first in any global contest in his profession, such as the World Agriculture Prize, World Soil Prize, Japan Prize, U.S. Awashthi IFFCO Award,

DOI: 10.1079/9781836990154.0003

World Food Prize, Arrell Global Food Innovation Award, Padma Shri from the President of India, and Gulbenkian Prize for Humanity from Dr Angela Merkel, among others. It is this confidence and 'can do' attitude which has been the key factor. Furthermore, it was also Rattan Lal's cultural immersion in village life and the rural community that made him a better-trained future soil scientist.

3.2 Transitioning from a School to a College

Most college students in India and elsewhere during the 1950s and 1960s were focused on obtaining education in medicine, law, engineering, business management, or basic sciences such as chemistry, physics, astronomy, geology, or petroleum engineering. Children from educated and well-to-do families had their goals in obtaining education in these subjects chosen by an elite segment of the society and followed the family tradition. However, Rattan Lal belonged to a special group of citizens who were farmers and practising the age- old tradition of growing crops and raising livestock as a way of life. Growing up under rural conditions in the 1950s (e.g. villages without roads, running water nor electricity, or other basic necessities of life), Rattan Lal became interested in studying agriculture, the profession that his family and the community practised over millennia. He could only think of the living conditions in which he grew up, with meagre crop yields (400–500 kg of wheat per acre or 200 kg of beans or chickpeas per acre) which were the best they could harvest under subsistence farming practices. He could only think of droughts, water scarcity, heatwaves, dust storms, invasion by desert locusts and soils which set hard and had to be ploughed with bullock-pulled *desi hal* or *ard* and meagre crop yields. He was focused on soil in which the plants grew, absorbed their water and nutrients, and produced some grain for people and stover for cattle. During the late 1950s and early 1960s, India was a food-importing country, strongly dependent on external help to obtain the food staple of wheat, rice, cooking oil and pulses. The nation was surviving on a ship-to-mouth scenario and this was the serious issue facing India and other countries in South Asia. During the 1950s and 1960s, food insecurity was the topic of discussion everywhere including newspapers, radio, civic discussions, college debates, political meetings etc. He could feel an atmosphere of helplessness in all walks of life. Thus, for a teenager (15 years old), graduate from a village school and brought up under rural agricultural settings, agriculture was the most important topic in his thoughts, and top of the agenda for any current and future plans.

Growing up in a farming family with small landholding, Rattan Lal experienced first-hand the constraints and challenges faced by resource-poor farmers – drudgery and hardship of ploughing the field with bullocks during the months of June and July when the temperature in the shade would reach 45°C, drought and scarcity of water, depleting fertility and degrading soil, low crop yield of poor nutritional quality, and little if any cash to purchase some basic needs (clothes, shoes, medicines). Motivated and driven by the desire to change this hardship, Rattan Lal focused on learning soil functions and how can these be restored by adopting practices of sustainable management of soil and crops, understanding soil restorative technologies, learning methods of conserving green water in the root zone, using systems or techniques to decrease soil temperatures during the summer, and reducing risks of crop failure so that no one suffered from hunger or malnutrition. Rattan Lal was committed to his roots and to developing farmer-friendly techniques of soil management for producing an adequate amount of safe, healthy and nutritious food while also restoring the environment so that there were no dust storms or flood/drought cycles.

3.3 Transitioning from Village to City Life

Rattan Lal had important family contacts in Ludhiana, a city where the new College of Agriculture was being established. Thus, after graduating from the village school, Rattan Lal arrived in Ludhiana to stay with his father's elder sister. Her family welcomed him, and the home of her aunt's family also became Rattan Lal's home for the next 4 years from July 1959 to July 1963. Some of the available amenities in the city of Ludhiana provided a unique experience for

Rattan Lal, e.g. street lights and municipal water taps. Water was available for specific times of the day (such as 6–9 am, and 4–8 pm). Residents of the community lined up for the water and carried it in buckets to their home. It was the only source of water for everything (drinking, cooking, bathing, washing clothes and kitchenwares, etc.). Scarce as the water supply in the city was, it was a step higher than the village water supply from a 30–40 m deep well from which it had to be lifted either manually or by animal-pulled devices. Some homes also had a hand pump, and others had the municipal water supply pipe connected to their homes. Then there were shopping facilities such as the Chaura Bazar, Pindi Street, for medicine shops and pharmaceuticals, Sabzi Mandi for fruits and vegetables etc. There was also a 'Committee Park' with nice flower beds and walkways. Above all, there were reading rooms where the newspapers were available, and public libraries. This was a monumental change for Rattan Lal and he availed himself of the library facilities and read books on soil science, agriculture and other themes including some on poetry.

3.4 Studying at an Agricultural College

Once the historic, complex, lengthy and challenging process of admission to the newly founded agricultural college was completed (see Chapter 2), and Rattan Lal was able to join classes about one month later than when the first-year class started, he noticed some differences between his village school and the city college. Indeed, the building was impressive despite still being under construction. Everything was new – the floor, the ceilings, classrooms with fans, seating, and the big auditorium with a gallery. Of course, research-cum-teaching laboratories were impressive such as that for zoology, entomology, pedology, botany, and others. The agronomy farm was still the old one and about 5 km (3 miles) away to the east and towards the city of Ludhiana. Surroundings of the buildings were still sand dunes and thorny shrubs criss-crossed by footpaths towards other rapidly urbanizing parts of the city. These educational facilities were high-level compared with basic rooms without seating and electricity of his village school.

Of all the subjects being taught during the first year, botany was the most interesting to him because it provided insights into the complexity of a seed and the plant that grows from it, of the process of photosynthesis driven by chlorophyll and solar energy, along with the uptake of water and plant nutrients from the soil. Agronomy was interesting and the practice of ploughing with bullock-driven plough was, although not new to him, intriguing. He often wondered, though, about the need for teaching how to use a bullock-pulled plough. It was certainly a novelty for his class fellows who came from urban backgrounds. Rattan Lal often thought about the need for a class dedicated to management of soil for growing crops in which ploughing, or eliminating it, should be a study. These thoughts were also strengthened by comments made by his village community who asked him on occasional return to his native village during the holidays, 'Why did you have to go to Ludhiana to learn how to plough? We could have taught you that right here in our own village. This rather humorous commnet haunted him for most of his professional career in which he tried to find out the reasons why any ploughing was needed.

While botany was interesting, Rattan Lal also learned that plants have to grow into a media to be able to flourish and produce grains and other edible parts. Indeed, the health of soil determines the vigour, growth and yield of a plant. Watching the Ram Leela, a drama organized in every village to celebrate the life of Lord Rama and the birth of his would-be wife (Sita or Siyya) from the furrow when King Janak was ploughing the land was a great learning experience. These and other mythological stories which he learned during childhood in his village were the foundation which kept the word 'soil' (*bhumi, dharti, dharti mata, dharra, vasundhra, khak*, etc.) in his thoughts. During the fourth year of his college (1962), Rattan finally made up his mind that he would major in soil science and learn how to manage it.

3.5 College Life

In comparison to village life, life in college was even better, and it was the foundation stone of his career. There were improvements in living conditions, and thanks to his bua ji, he was very well looked after in terms of providing him with basic necessities. While the walk to the college was a challenge, because it was 4 miles (6 km) away and there was no bicycle, it did not matter. The entire environment was new and invigorating and provided a conducive environment in which to learn. Every day at the college brought a new opportunity for him to learn more about soil science, agronomy and other disciplines related to agriculture and environment.

But, there were challenges. The most significant part was getting admission to the Government Agricultural College and Research Institute. He dreamed about being a part of the college under construction but the admission was neither a simple nor a pleasant process. It was the biggest challenge to be faced by a 15-year-old youth to realize his dream of studying soil, agronomy and the environment of agro-ecosystems. Being denied admission – a very big blow – had to be reversed but he did not know how to meet this challenge.

It is said 'Where there is a will there is a way'. Rattan Lal found the way to study the subject of his choice: protection, restoration and sustainable management of soil to address global issues.

Bibliography

Khan, N. (2024) Village life: A deep dive into community, tradition, and sustainability. *Medium*. Available at: https://nabeel-khan.medium.com/village-life-a-deep-dive-into-community-tradition-and-sustainability-82ad318df17f (accessed 28 May 2025).

4 Return of the Native to His Home after Four Years of Living in the Western World

'I would rather live on these hills than anywhere else in the world.'
'The sea changed, the field changed, the rivers, the villages, and the people changed, yet Egdon remained.'
— Thomas Hardy, *The Return of the Native*

Abstract

Rattan Lal's interest in smallholder farming based on his rural community life was strengthened by his appointment as a soil physicist at IITA in Ibadan, Nigeria. While his one-year of stay at the University of Sydney was a great experience, he wanted to devote more time to addressing practical problems of real-world agriculture than conducting controlled laboratory experiments. His visit to his home in Rajound and to his rural community en route from Sydney to Ibadan was also a motivational force emphasizing the need to support small landholder farmers of the world. Rattan Lal's drive from Lagos airport to the city of Ibadan through 90 miles of rural roads in 1969 was a unique experience. He was fascinated by the scenic drive through the dense tropical rainforest ecosystem of the humid tropics, and looking at the traditional food crops in the village market en route. These tropical ecosystems and the food staples he had neither seen before nor even studied in the textbooks. The very first soil erosion of a ploughed field by a tropical rainstorm was an eye-opener for him that he could not forget for the rest of his professional life. Therefore, he was motivated and determined to find viable alternatives to shifting cultivation and bush fallow systems of the tropics.

4.1 Interest in Serving Small Landholder Farmers

During December 1968, the same month that he had arrived in Sydney, Rattan had received a letter from Dr Herb Albrecht, whom he had met in the office of Dr Garth Volk at The Ohio State University. Dr Albrecht was now the Director General of IITA and was based in Ibadan, Nigeria. Dr Albrecht offered Rattan Lal a position of soil physicist at IITA but indicated that there was no rush to take up the position immediately because Nigeria was then under a civil war (the war of Biafra). Dr Albrecht also indicated that Rattan Lal's mission at IITA would be to 'increase quantity and quality of food production in sub-Saharan Africa, and identify viable and sustainable alternatives to shifting cultivation'. Rattan Lal did not disclose this letter of offer to Prof. Collis-George for several months. When he finally plucked up the courage to inform Prof. Collis-George about his decision to leave Sydney and go to Nigeria, the professor was rather surprised, and asked, 'Did I hear correctly that you plan to leave Australia and go to Nigeria?' When he replied in the affirmative, the professor asked him what research in soil physics he would be doing in Nigeria. Rattan Lal replied that his task would be to 'improve the quality and quantity of food production in Africa'. Professor Collis-George was amazed and surprised at Rattan Lal 's decision to accept such an ambitious and aspirational job, and to leave Australia for a country with a raging civil war. Professor Collis-George asked Rattan Lal how long this task of improving the quality and quantity of food production would take, and how knowledge of soil physics would be of relevance to achieving such a task. While Rattan Lal had no answer to these questions, he had made up his mind that his future lay

DOI: 10.1079/9781836990154.0004

in serving the small landholders of Africa, Asia and elsewhere. Yet Prof. Collis-George was right about the lack of wisdom of a 25-year-old person to leave Sydney and to live and work in Nigeria. Rattan Lal remembered the seriousness of this decision on the very first day of his arrival in Nigeria when he undertook the memorable road journey of about 90 miles from Lagos to Ibadan through countryside and narrow and pot-filled roads. (The modern Lagos–Ibadan highway was constructed in the early 1980s.)

4.2 The Homecoming to Rajound

Rattan Lal departed from Sydney in early December 1969 and visited his family in Rajound. The KLM flight booked by IITA from Sydney to Lagos via New Delhi was first-class. Despite being a vegetarian and a teetotaler, travelling in first class was a unique experience. This was the first time he had travelled in first class. On his arrival at the airport in New Delhi, his brother and brother-in-law received him. They had made an arrangement with some family acquaintance in Old Delhi and arrived at their home early in the morning. The family was very kind and welcomed Rattan Lal to their home. However, he felt very uncomfortable and decided to visit his friends and class fellows at IARI. He arrived at the CCA (Central College of Agriculture) hostel and visited some of them during the morning. Later, he returned to the family and went to the railway station to take a train from Delhi to Karnal. From Karnal, three of them took a bus ride for about 40 miles (65 km) to Rajound. They arrived in Rajound at about 4 pm in the afternoon.

The journey in the first-class cabin turned out to be a strong contrast to the first night stay in Rattan Lal's childhood home to which he had returned after 4 years in Columbus, Ohio, and Sydney, Australia. There were winter rains at the time of his arrival, and the mud roof of the home lit by a kerosene lamp was leaking in several places. The fire in the middle of the room, with fuelwood and other biomass, kept the room warm but also filled it with smoke. During the night, his father and brother kept shifting Rattan Lal's homemade charpai (bed) from one place to another to avoid water dripping from the roof. Yet it was pleasant to catch up with the family events over the previous 4 years.

The following morning, Rattan Lal asked his father to come with him to the market town of Kaithal (25 km north-west). It was a pleasant journey on a crowded country bus with standing room only and some passengers sitting on the roof. They both went to a bank and Rattan Lal deposited US$ 4000 (Rs 30,000 at the time) in an account under his father's name, and asked him to use this money to build a new home. His father, pleasantly surprised with the gift, responded that he will do so but the home will be of a design similar to that which he had left behind in Pakistan. Rattan Lal agreed with that suggestion. Both returned to Rajound and discussed the design of the new family house on the way.

In addition to the house, there were some other issues that Rattan Lal had to address. His elder brother had a back problem and even a short walk was causing some severe pain. Thus, Rattan Lal spent several days to take his brother to a doctor in Karnal and eventually to a hospital in New Delhi. It turned out that treatment with some vitamins and micronutrients was all that was needed.

Rattan Lal also visited his college at Ludhiana and met some of his professors. During his visit to PAU, Prof. George Taylor was also visiting. This was a pleasant coincidence. Rattan Lal discussed with Prof. Taylor his plans for the new job in Nigeria.

As promised by his father, Rattan Lal returned to the newly built home during the next visit for a one-month vacation from IITA in December 1970. The new home had electricity, and the front yard had a hand pump for water. Rattan Lal also gave his father another US$10,000 to purchase farmland similar to that which he had in Pakistan. In 1971, his father purchased 12 acres of farmland, which was equipped for irrigation by a canal. His father was now the proud owner of a home and farmland exactly 25 years after leaving this family heritage behind in Pakistan. He lived in the house and farmed the land until he died in 1978. Any time Rattan Lal visited his village between 1970 and 1978 and informed his family that he was arriving on a specific day, his father would walk about two miles to a bridge on an irrigation canal and wait for Rattan Lal's taxi to arrive from Delhi.

Rattan Lal asked his father not to put his name on the house and the land. The house and land were now owned by Rattan Lal's elder brother and his children. Later, after the death of his father in 1978 and of his elder brother in 2010, Rattan realized that he had no connection whatsoever with his village family and with the native village community where he studied.

Looking back, his decision not to put his name on the house and land he helped purchase was shortsighted. By the 1990s to early 2000s, he realized that he had no contacts at all with his native village. After his father and brother passed away, Rattan did not know his nephews and nieces and lost touch with his village community. Both socially and culturally, this was a serious mistake. He lost touch with the very foundation of his family heritage and community life.

Whenever Rattan returned from IITA to his village, he always rented a car from New Delhi airport. At the time, the car rental (with driver) was Rs 40 per day plus the cost of fuel. Drivers of most cars he rented informed him that the speedometer in the car was not functional. Thus, whenever the driver went away for lunch or any other event, there was no record of the mileage. In December 1970, when Rattan Lal returned from IITA for the first time, the car broke down about ten miles from Rajound. There was no transport; thus, he rode the remaining distance on a mule cart. These rental taxis were not self-starting. Every morning, when he was at the village with his taxi, it had to be pushed for several yards before it would start. Many young men were pleased to volunteer to push-start his taxi.

The trip to Rajound was always adventurous and pleasantly memorable. Whenever, Rattan Lal visited his home on the annual vacation from IITA, his village community always asked him questions about his life in Nigeria, how the people of the country were, what African villages look like, what language they spoke, what kind of food they ate, and what their social customs and traditions were. They were fascinated to learn that food staples (rather than wheat) were yam, cassava and other root crops, and their social customs (e.g. polygamy) were also different.

4.3 The Drive through a Tropical Rainforest

Rattan Lal departed New Delhi on 27 December 1969 via Beirut and Accra to Lagos, Nigeria. At the Palam airport in New Delhi, Rattan tried to board the plane via stairs at the front part of the plane for the first-class cabin. However, assuming that he could not be a first-class passenger, the ground staff advised him to board the plane at the rear and mentioned that this entrance was for first-class passengers only. Upon checking his boarding card and passport, after the other passengers had already boarded, Rattan Lal was finally allowed to board the plane at the front.

After staying overnight at the IITA guest house near Lagos airport, he was driven straight north to Ibadan. The long journey on a winding narrow road was frequently interrupted (every 15–20 km) by the army check-posts, which took a lot of time. Despite being nervous due to frequent army checks, which resulted in scattering of his meagre belongings in the baggage compartment, Rattan Lal was fascinated by the scenic drive through the dense tropical rainforest ecosystem of the humid tropics that he had neither seen before nor even studied in textbooks. The driver was courteous and polite, and besides reassuring the frightened passenger, he also pointed out some trees in the forest: cocoa (*Theobroma cacao*) whose beans are made into chocolate, oil palm (Elaeis Guineensis) whose seeds are used to extract the oil for cooking food, and cola (*Cola nitida*) from whose seed is the source of flavouring for the famous Coca-Cola drink. The driver also pointed to a shrub which was obviously a cousin of banana and fruit of plantain (*Plantago major*), similar to that of banana but which needs to baked or roasted prior to eating. In a village market where the driver stopped, the driver also showed Rattan Lal tubers of cassava (*Manihot esculenta*) and yam (*Dioscorea rotundata*), which have been food staples in West Africa but hitherto completely unknown to Rattan Lal. At a crowded truck stop, midway to Ibadan, the local market had a circular charcoal-based open fire for roasting shish kebab, which was served with baked plantains wrapped in banana leaves. Local shops and trolleys also carried mango, guava, papaya and some leafy vegetables (amaranth). There were containers containing grains of sorghum, millet and of the flour made from fermented cassava tubers (gari). Cassava grown in Africa was a bitter cassava, and the cyanide is removed by the process of fermentation before it can be eaten. Some of the crops that Rattan Lal was familiar with (wheat, barley, canola, chickpea, sesame, lentil, mung bean, urd beans) were nowhere to be seen. At the time, even rice was not a common food grain, but sugarcane stalks were available in some shops. The drive, tiring and full of anxiety as it was, turned out to be a refresher course in vegetation, physiography and landscape of a humid tropical ecosystem and staple food crops of sub-Saharan Africa. Textbooks and classes taken at PAU, IARI, Ohio State and the University of Sydney had not prepared Rattan Lal for such a fascinating biome. The six-hour journey was indeed an open textbook under the tropical sun.

Finally, at sunset, the driver indicated that they had arrived on the outskirts of Ibadan. It was a sprawling coalition of unending homes, with rusted tin roofs and winding streets, with a population of more than 1 million in December 1969. Towards the extreme north of the town, on a hill, the driver stopped in front of the Premier Hotel. In the hotel lobby, Rattan Lal met Dr Albrecht, whom he had met about 15 months before in the office of Dr Garth Volk at the Ohio State University. Dr Frank Moorman, senior soil scientist, who had joined IITA about six months previously, was also present. When asked by Albrecht and Moorman how the journey was, Rattan Lal replied, honestly and sincerely, that he would like to go back to Sydney, if possible by a return flight the next day. It was a response that the doctors were not expecting. During the first night in Ibadan, Rattan Lal remembered the surprised look of Prof. Collis-George who wanted to be sure that he had heard correctly Lal's decision to leave Sydney and go to Ibadan, Nigeria. Now that he was in Ibadan, Rattan Lal wished to be in Sydney, Columbus, New Delhi or even in the remote village of Rajound. Miraculously, and against all odds, however, Rattan Lal stayed at IITA for almost 18 years from 28 December 1969 to 10 July 1987.

4.4 Arrival at the International Institute of Tropical Agriculture, Ibadan, Nigeria

Rattan Lal arrived at IITA on 28 December 1969, and it was the dry season. The sky had a milky colour because of the Harmattan dust, and the mornings occasionally had fogs. Most of the morning sky had low clouds. Indeed, the total radiation and sunshine hours could be less in the humid tropics than those in Alaska. Sunshine was observed late in the morning (usually after 11 am) and the hottest time of the day was about 2 pm.

Ibadan, in the West African region, about 6°N latitude, had a bimodal rainfall distribution. There were two rainy seasons: the first, longer, growing season from the end of March to the end of July and the shorter second season from September through mid-November. These two rainy seasons were separated by a shorter dry spell in August and a longer dry season from mid-November to the end of March. The end of the second rainy season was brought about by the dry winds blowing from north to south and carrying with them the Harmattan dust, which imparts the milky colour to the sky. Harmattan, a cool dry wind that blows from the north-east or east in the western Sahara, was strongest from late November to mid-March. It usually carried large amounts of dust, which was transported hundreds of kilometres over the Atlantic Ocean; and the dust often interfered with local air transport operations. Harmattan had strong ecological effects on the West African region (Jenik and Hall, 1966).

The fine dust of Harmattan could be detected in the Caribbean in the west and in Scandinavian countries in the north. It was during the dry season that soil physical properties could be measured under field conditions. Pertinent soil physical properties to be measured *in situ* under field conditions included bulk density, water infiltration rate, soil moisture profile, soil temperature regime and its diurnal fluctuations, etc.

One day, in March, Rattan Lal asked his field staff when they could expect the rain to begin. A member of staff replied that 'rain will not happen until the house of my village priest being constructed is completed'. When Rattan Lal sought clarification about priests' powers to stop and start the rain, they were surprised that the young researcher did not know about priests' magical powers, having direct communication with the 'rain god'. This incident reminded Rattan Lal of similar customs in his native village when the priest would organize a community worship (*havan*) as prayers to Lord Indra (the rain god of Hinduism) to start the onset of monsoons.

Towards the east of Ibadan, about 15 km from the Premier Hotel and in close proximity to the campus of the University of Ibadan, was the 1000-acre site of IITA. It had a 2.5 m high barbed-wire fence and its entrance was guarded by tight security. Among buildings under construction were the farm building, 300, and those numbered 400 and 401 for crops and farming systems research, respectively. Soil science was part of the farming systems. The library, administration building 500, auditorium and the international house were in close proximity. Next to the designated area for farmland was the garage for vehicle repairs along with carpentry and metal workshops, and a gasoline pump. There was a deep pit next to the international house, and it was for a swimming pool. This pit also served as an example of the first soil profile examined by some soil scientists of IITA. A small operational building on site, used by construction engineers and painted white, was popularly known as the White House. Once the construction workers vacated it, the kitchen in this building served as the first laboratory Rattan Lal worked in at IITA for the initial year until the laboratory buildings were completed.

The scientific staff of IITA, about eight in total by January 1970, comprised 3 soil scientists (Moorman, Kang and Lal), 4 plant scientists (2 breeders, 1 plant pathologist and 1 entomologist) and an economist. All scientists were provided desk space in a house in a suburb of Ibadan. Bedrooms were the offices, two or three desks per room; the kitchen was the laboratory; and living room with a long table with 12 chairs was the meeting room. Weekly faculty meetings were chaired by the Associate Director (Dr Rodney Briggs). During one of the meetings in January 1970, the Associate Director announced that IITA was to be officially inaugurated by General Gowon, Head of State of Nigeria, on 21 April. Thus, some land needed to be prepared and field experiments established to show progress in research.

4.5 *In Situ* Measurements of Soil Physical Properties under Forest Cover

The farm manager assigned Rattan Lal four or six field workers, and each was equipped with a machete and a short-handled hoe. The farm manager also mentioned that more field workers could be assigned if needed. Most of these workers were primary-school graduates and one of them was a high-school graduate and spoke good English. From January through March, with the help of these workers, Rattan Lal made some in-field measurements under rainforest on soil bulk density (core or excavation method), water infiltration rate (double-ring infiltrometer), and penetration resistance (pocket penetrometer). Rattan Lal also purchased a few mercury-in-glass thermometers to measure soil temperature under forest and on clear, bare ground surface. While the bulk density was low (0.9–1.1 Mg/m^3 of the surface layer) and water infiltration rate was high (10–30 cm/hr) in soil under forest, the bulk density was high (1.3–1.4 Mg/m^3) and infiltration rate low (2–5 cm/hr) in soil after the land was cleared of its vegetation. An important difference was in the soil temperature at 2 cm depth which was almost constant at about 25–30°C under forest compared with 45–50°C under bare soil at 2 pm on a sunny afternoon. The high soil temperature reminded Rattan Lal of his cowboy days in Rajound when he and his cattle longed to be under the cool and refreshing shade of a tree.

These field workers had farm backgrounds, and Rattan Lal learned many practical things by working with them. Some of these traditional practices were not written in the college textbooks. Rattan Lal wished that he understood their language and could talk directly to each one of them.

There occurred an exciting incident when Rattan Lal was with one of his field workers and getting ready to make measurements on soil bulk density under a forest canopy. The worker did not speak English. When Rattan Lal was getting ready to clear the detritus material from the soil surface, the worker tried to say something to get his attention. When Rattan Lal neither moved nor responded to the staff, the field worker pushed Rattan Lal, who fell on the ground a few metres away. Despite being frightened and confused, Rattan Lal stayed calm, kept lying on the ground and thought about the next move of the worker. However, the worker kept pointing at the pile of leaves and branches where Rattan Lal was standing before and made hand signals for him to keep away. The worker then cut a tree branch of 3–4 cm diameter and about 1–1.5 m long, sharpened its edge 4–6 cm long with a pointed tip, and pushed the sharpened edge into something on the ground that Rattan Lal had not yet seen. With both hands, the worker picked it up by the head pierced by the sharpened branch. It was the Gbong viper, 1.5–2 m long and 20 cm in circumference. With its mouth wide open and fangs visible, it was alive. By pushing the peg further into the ground and making sure that the snake was securely pinned, the worker cut a twining vine 3–4 m long and tied the snake into a bundle with the stick pierced through its head. The securely tied bundle was put in the back of the truck and they drove to the farm building on the main campus of IITA. The tied-up snake attracted quite a crowd. To calm it down, the snake was put into a bag and transferred into a cold-storage room for the night, and the worker claimed it the following morning. The head worker explained to Rattan Lal that this was the way snakes were hunted. Frightened and confused as Rattan Lal was when being pushed away in the forest, he was grateful to the worker who saved him from the danger and enabled him to measure bulk density some other day.

When field runoff plots were established to monitor soil erosion, snakes would also wash into the water-storage tanks. For both scientific and personal reasons, it was a good idea to stir the water in the tank with a stick prior to dipping a bottle in it. Spitting cobras were also dangerous. Cover crops (e.g. *Mucuna utilis*), which created a dense shade and cooler soil beneath their canopy, attracted snakes. Thus, wearing knee-length boots and using glasses to safeguard against spitting cobras were essential field gear. Green mambas were considered the most venomous and difficult to see among green foliage. Wearing long rubber boosts was essential to walk through rice paddies because swamps were infested with snails which transmitted bilharzia, mosquitoes which caused cerebral malaria, and organisms which caused swellings of feet, and river blindness. Every time Rattan Lal went to the field plots for conducting some soil measurements, he asked for the same worker to accompany him.

Another example of the traditional knowledge of these farm workers was about termite control. Rattan Lal was using a plastic sheet to line a trench to store runoff water from an experimental plot. However, the plastic was being eaten by termites within a few days and insecticides were rather ineffective. A worker suggested using the ash obtained from burning branches of a specific tree before putting the plastic on the ground. This ash was an effective termite repellent.

4.6 Soil Erosion by a Tropical Rainstorm

The farm manager cleared some land, installed terraces to control runoff, ploughed the land on the contour (across the slope) with a disc plough, and made a smooth seedbed. Along with the plant-breeding plots, Rattan Lal established an experiment to study the effects of methods of seedbed preparation and time of planting on maize germination, growth and yield. Treatments involved ploughed bare soil; ploughed with crop residue mulch; ploughed with ridges across (perpendicular) to the slope; and mounds about 30 cm high and 50 cm in circumference. Yams grown on top of mounds were traditionally intercropped with maize (maize) or cowpea. Rattan Lal planned to monitor soil temperature, soil moisture content and other parameters *in situ* such as penetration resistance, water infiltration rate in the furrow and ridge top, etc.

The evening before the inaugural ceremony and visit of the Board of Trustees, there was a rainstorm of about 10 cm that fell for the first 15 min at the rate of 15 cm/hr. The next morning when Rattan Lal arrived at the field, he was awestruck because all plots, except the one which were unploughed or were covered with the mulch, were washed away and seedlings were carried into the terrace where the water runoff was ponded (Fig. 4.1). The terraced dyke was breached at several places and washed away. Rattan Lal had never seen such devastating soil erosion before and informed the farm manager not to bring the Board of Trustees to visit the experimental plots.

A member of the Board of Trustees was Dr Dennis J. Greenland (from now on DJG), whom Rattan Lal had met about 1 year previously at the University of Sydney. DJG, who had since returned to Reading as the Head of the Soils

Fig. 4.1. Severe soil erosion under maize seeded with conventional plough-based method of seedbed preparation caused by a rainstorm experienced at IITA, Ibadan, on 20 April 1970, which fell at the intensity of 15 cm/hr for the first 15 min. Compared with this, the unploughed soil and the plots covered with crop residue much did not have any runoff nor any erosion. Photo: Rattan Lal personal photo album from IITA.

Department, was surprised to see Rattan Lal working in the humid tropics, clearing land and studying the impact of deforestation and ploughing on soil physical properties and processes. Rattan Lal heard DJG mentioning to Dr Albrecht the drastic transformation in a young soil scientist. DJG knew Rattan Lal as someone researching thermodynamics of water movement in a vertisol under sophisticated laboratory conditions using tracer techniques and time-lapse photography. Just 1 year later, the laboratory scientist was working under field conditions to study the effects of tropical deforestation and of ploughing on soil erosion and changes in microclimate.

DJG had a long discussion with Rattan Lal and encouraged him to focus on monitoring soil physical properties (e.g. structure, crusting, compaction, soil temperature, water retention and movement, and runoff and erosion) upon conversion from forest to agricultural ecosystems, and understand the cause–effect relationship. Having written the classic book *Soils under Shifting Cultivation*, DJG was the guru of sustainable management of soils of the tropics in general and of sub-Saharan Africa (SSA) in particular.

Because of his interests in soil management and farming systems, DJG was appointed Deputy Director General of IITA. DJG joined IITA in 1974, which created a unique opportunity for Rattan Lal to work with a great soil scientist of his time. Together, they organized conferences on 'Soil and Water Conservation in the Humid Tropics' in 1975 and 'Physical Properties of Soils of the Tropics' in 1977 (Greenland and Lal, 1978; Lal and Greenland, 1979). When DJG left IITA in 1978, Rattan Lal organized a third conference on 'Deforestation and Tropical Hydrology' in co-operation with Dr Walter Russell, also from Reading and author of the famous book *Soil Conditions and Plant Growth* (Lal and Russell, 1981). Rattan Lal also organized a conference on tropical deforestation and published a book (Lal *et al.*, 1986).

4.7 Dawn of No-Till Farming in Tropical Africa

Since the experience of seeing a spectacular soil erosion in April 1970, Rattan established a series of field runoff plots on different slopes to study effects of slope, soil management and crop rotations on runoff and erosion (Fig. 4.1). He also established long-term no-till (NT) experiments to study the effects on soil properties and crop growth (Fig. 4.2). The data showed that runoff and erosion were effectively controlled by residue mulch of 4–6 t/ha on a ploughed soil, and by use of NT with the residue retention. These experiments documented that soil erosion was caused not by what crop was grown (open row vs closed-canopy crop) but by how it was grown (NT vs plough till). Soil temperature and moisture regimes were favourable under NT and earthworms flourished under NT (Fig. 4.3).

Soil-moisture content and soil temperature were more favourable under no-till than under plough-based system. The data from runoff plots and watershed management showed that no-till system effectively controlled water runoff and minimized the risks of soil erosion up to slope gradients of 15% in these highly erodible soils and erosive rainfall events. Rattan Lal started no-till experiments at IITA in 1970, and thus may be the first scientist to begin research on to-till farming in the tropics (Fig. 4.4).

Based on promising results on field plots, DJG suggested that Rattan Lal conduct NT studies on large watersheds equipped with flume, water-stage recorder and sampling devices. Thus, paired watersheds of NT and ploughed treatments were established on sloping land of 4–5 ha each. Effects on soil properties, water runoff and erosion were studied for about 5 years. Soil under NT watershed was prone to compaction, and yet erosion was effectively controlled. Crop yield began to decline after 3 years because of soil compaction caused by harvesting with combine in August when the soil was wet (Figs 4.5, 4.6 and 4.7).

Fig. 4.2. Long-term field runoff plots were established at IITA on different slope gradients, and these were used as training facilities for scholars from around the humid tropics. Photo: Rattan Lal personal photo album from IITA.

Fig. 4.3. Soil under native forest and under no-till method of seedbed preparation had intense activities of earthworms which increased water infiltration rate and decreased risks of soil erosion. Earthworm casts were also enriched in plant nutrients compared to the soil from which they were derived. Photo: Rattan Lal personal photo album from IITA.

Fig. 4.4. Cowpeas grown through the residues of maize by no-till farming at IITA during the second growing season of 1971.

Both on NT and plough till watershed, there was also a problem with root nematodes. At the advice of a nematologist, the farm manager applied Furadan at the rate of 2 kg a.i./ha. Within a few days, soil structure collapsed, all macrofauna (earthworms, termites, centipedes, millipedes, and presumably the microbes, were dead). The soil was obviously sterilized, it was a dead soil. Adding topsoil from a nearby forest did not recover it for one or 2 years.

Fig. 4.5. Undisturbed monolith weighing lysimeters were established within the watershed to study water balance and leaching losses at IITA in 1977. Photo: Rattan Lal personal photo album from IITA.

Long-term experiments conducted at IITA and other sites in West Africa showed promising results providing that crop residues mulch was left on the soil surface and a cover crop was grown at the end of the rainy season. Removal of crop residue mulch had adverse effects on maize growth and yield at IITA (Fig. 4.6) as it did at the USDA Experiment Station in Coshocton, Ohio (Fig. 4.7). Establishment of a cover crop, at the end of the growing season (from early November to early April), would maintain the ground cover, enhance soil fertility and reduce risks of soil erosion during the onset of the rains in April (compare Figs 4.1, 4.3 and 4.8).

Whereas the NT experiments at IITA in Ibadan were established since August 1970, and continued for 18 years till 1987, there was virtually no adoption by small landholder farmers of Nigeria or SSA. Part of the cause of lack of adoption of NT was the contradictory recommendations of other institutions working in sub-Saharan Africa (SSA), and partly due to competing uses of crop residues.

4.8 Scientific Debate on Concepts of No-till Farming

When DJG was the Associate Director of IITA, the Chairman of the Board of Trustees was the Director of IRAT, the agricultural research institute of France. Founded in 1946, The Institut National de la Recherche Agronomique was a French public research institute dedicated to agricultural science. IRAT had excellent field studies in each Francophone country of Africa (e.g. Senegal, Ivory Coast, Togo, Benin, Burkina Faso, Niger, Cameroon and those in northern Africa). IRAT scientists had data from long-term tillage experiments and its recommendation was mould board ploughing at the end of the rains (October) and rotary tillage at the beginning of the rains (April) followed by seeding of sorghum, maize or millet. Thus, the Chair of the Board (DG of IRAT) was surprised by Rattan Lal's recommendation of NT farming with the crop residue mulch. Both Dr Albrecht and DJG were aware of these contrasting results and encouraged Rattan Lal to contact researchers from IRAT

Fig. 4.6. Effects of continuous removal of crop residues from the plot in the foreground compared with residue retention on the plot in the background shows the benefits of residue mulch on soil functions. These experiments, conducted at IITA by the late Dr A.S.R. Juo and Dr Rattan Lal, were photographed after about 15 years of residue removal which showed severe adverse effects on soil properties and thus on maize growth and yield. Photo: Rattan Lal personal photo album from IITA.

and visit their experiments in the neighbouring countries.

Rattan Lal visited experiments of IRAT in Senegal and Ivory Coast and was very impressed with the high quality and systematic research done by IRAT's agronomists. Rattan Lal learned many new ideas from discussions with IRAT researchers, who were extremely co-operative and friendly. However, they shared with him the data on consistently poor performance of NT plots on crop yield compared with those of the ploughed plots. While visiting the plots, low crop height and poor growth of crops in the NT plots were visible even from a distance. On a closer look on the soil surface, Rattan Lal found that soil in NT plots was bare, crusted, compacted and very hard. When asked why there is no mulch cover, Rattan Lal was informed that the crop residue was removed because this was what the farmers did. Rattan Lal then realized that what IRAT scientists called the NT was not what he had been practisng at his plots at IITA. Indeed, Rattan Lal practised NT with residue retention and complex rotations, and IRAT scientists eliminated tillage as well as removed crop residues. Removal of residues made soils prone to accelerated erosion by water and wind, crusting and compaction, and depletion of plant nutrients because of disruptions in nutrient cycling. The field visit followed by discussions in the laboratory during the mid-1970s were witnessed by a newly arrived postdoc (Dr Christian Feller) from IRAT. Dr Feller recalled these differences in definition of NT farming during an event in 2019 in Tokyo when Rattan Lal received the Japan Prize. While the debate on this topic continues even during the 2020s, several attempts have since been made to define a system-based conservation agriculture (CA) with key pillars. FAO (2016) defined three pillars of CA comprising no ploughing, residue retention, and complex crop rotations. Lal (2015) defined CA with five pillars: no ploughing, residue retention, cover cropping in the off-season, integrated soil fertility management,

Fig. 4.7. Continuous removal of crop residues had the same adverse effects on a soil at Coshocton, Ohio, as that in Ibadan, Nigeria. The person in the photo is Dr Josh Beniston who conducted his thesis research at the USDA Experiment Station at Coshocton. (Photo: USDA ARS Coshocton, Ohio).

and integration of crops with trees and livestock. Other variants of these system-based approaches to CA include combination of a mulch/cover crop-based NT with sustainable intensification, and tree-based rotations. Continuous removal of crop residues from NT plots adversely affected agronomic growth and yield because of severe degradation of soil properties and processes (Fig. 4.6). The plot in the foreground of Fig. 4.6 shows the effects of residue removal for 15 consecutive years. The plot in the background has everything else the same as that in the foreground but the residue mulch was retained on the soil surface after harvest of the crop. Indeed, at the time of taking this photo in 1987, soil in the foreground was a dead soil; it was devoid of its biotic activity, and thus, the surface layer was crusted, compacted and not responsive to input of chemical fertilizers. Removal of crop residues also caused severe depletion of its soil organic matter (SOM) content, which has been considered the heart of soil health. Thus, continuous removal of crop residues and use of inorganic fertilizers to restore soil fertility led to severe degradation of soil health. This is exactly the problem of soil health in sub-Saharan Africa.

The need for a widespread adoption of a system-based CA is greater during the 2020s than ever before. The world is on a crossroads faced with climate change, COVID and conflicts (3 Cs). Furthermore, Sustainable Development Goals (SDGs), similar to the Millennium Development Goals and Agenda 21, are not on track for accomplishment by 2030. The resource-poor and small landholders, estimated to be 550 million globally, neither have access to resources nor the institutional support needed for adoption of a system-based CA comprising of three or five pillars. Incentivization, through payments for strengthening of ecosystem services, is critical to the adoption of CA (see Chapter 15).

Limiting global warming to 2°C necessitates a widespread adoption of regenerative agriculture leading to a negative emission farming (Lal, 2020; Smith *et al.*, 2018). Elimination of hunger

Fig. 4.8. A cover crop of velvet beans (*Mucuna utilis*) in the watershed experiment at IITA growing on residual moisture at the end of the rainy season. These watershed management experiments were established on a 40-ha watershed in 1977 and monitored for ten years until 1987. Photo: Rattan Lal personal photo album from IITA.

(affecting 733 million in 2024) and hidden hunger or malnutrition (affecting 2–3 billion) require adoption of nutrition-sensitive agriculture. Thus, the assignment of Rattan Lal by Dr Albrecht to improve the quantity and quality of food production was pertinent even during the 2020s, albeit with global scope and greater urgency. The task assigned to Rattan Lal in the 1960s remains a work in progress even during the 2020s.

4.8.1 Viable alternatives to shifting cultivation and bush fallow rotation

While working at IITA, Rattan Lal discovered that a viable alternative to shifting cultivation was a system of soil and crop management with the following 12 characteristics (Lal, 1975, 1976; Lal and Greenland, 1979; Lal and Russell, 1981; Lal *et al.*, 1986):

1. Eliminate any physical disturbance of the soil.
2. Maintin soil cover with a crop residue mulch, and grow a cover crop.
3. Leave any roots and stubble in the soil and do not burn biomass.
4. Establish a cover crop during the off-season, which in the case of Ibadan was the long dry season from the end of October to mid-April.
5. Adopt complex crop rotations including integration of crops with trees and, where possible, with the livestock.
6. Strengthen nutrient recycling and use of biological nitrogen fixation.
7. Supplement soil fertility with judicious use of chemical fertilizers.
8. Observe the 'Law of Return' (Sir Albert Howard, 1943).
9. Follow the strategy of integrated nutrient and pest management.
10. Reduce postharvest losses of grains.
11. Create value addition of the farm produce.
12. Adopt regenerative agriculture.

These principles have been revised and updated during the subsequent chapters in the context of recent developments in soil science and managing soils beyond food and fuel.

Bibliography

FAO (2016) Conservation agriculture. Food and agriculture organization of the United Nations. Rome.

Greenland, D.J. and Lal, R. (1978) *Soil conservation and management in the humid tropics*. J. Wiley & Sons, London.

Hardy, T. (1878) *The Return of the Native*. Belgravia.

Howard, Sir A. (1943) *An Agricultural Testament*. Oxford University Press, London.

Jenik, J. and Hall, J.B. (1966) The ecological effects of the Harmattan wind In the Djebobo Massif (Togo Mountains, Ghana). *Journal of Ecology*, 54, 767–779.

Lal, R. (1975) Role of mulching techniques in tropical soil and water management. *IITA Technical Bulletin* 1.

Lal, R. (1976) Soil erosion problems on an alfisol in western Nigeria and their control. *IITA Monograph*, Vol. 1. Ibadan, Nigeria.

Lal, R. (2015) A system approach to conservation agriculture. *Journal of Soil and Water Conservation*, 70(4): 82A–88A.

Lal, R. (2020) Regenerative agriculture for food and climate. *Journal of Soil Water Conservation*. DOI: 10.2489/jswc.2020.0620A.

Lal, R. and Greenland, D.J. (1979) *Soil Physical Properties and Crop Production in the Tropics*. J. Wiley & Sons, London.

Lal, R. and Russell, E.W. (1981) *Tropical Agricultural Hydrology: Watershed Management and Land Use*. J. Wiley & Sons, London.

Lal, R., Sanchez, P.A. and Cummings, R.W., Jr (1986) *Land Clearing and Development in the Tropics*. Balkema Publishers, Rotterdam.

Nye, P.H. and Greenland, D.J. (1960) *The Soil under Shifting Cultivation*. Commonwealth Bureau of Soils, Farnham Royal, UK.

Smith, D.M., Scaife, A.A., Hawkins, E., Bilbao, R., Boer, G.J. *et al*. (2018) Predicted chance that global warming will temporarily exceed 1.5 C. *Geophysical Research Letters*, 45(21), 11–895. https://doi.org/10.1029/2018GL079362.

5 Soil, Ancient Civilizations and Modern Science

'Quicquid plantatur solo, solo cedit.' ('Whatever is affixed to the soil, belongs to the soil.')
— *Roman Maxim*

Abstract

Rattan Lal had interests in spiritualism and philosophy based on teachings in the Sanskrit/Vedic scriptures by his bebe. Thus, he often combined science with religious philosophy. He also looked at commonalities among the major religions of the world in the context of stewardship and management of soil and protection of natural resources. With the increase in scientific knowledge of agriculture and soil management, gods and goddesses of traditional religions were being slowly replaced by disciplines of soil sciences (pedology, soil chemistry, soil physics, soil ecology, soil microbiology, etc.) and those of plant sciences (plant physiology, plant breeding and genetics, and plant protection). Yet, Rattan tried to combine science and religion in creating the concept of soil, spirituality, stewardship and sustainability. He believed that the problem of feeding the growing and increasingly affluent human population necessitates addressing this major problem with all tools available to humanity. Based on the historic accomplishments of last 10 millennia, more changes may occur between 2020 and 2050 than have happened in the past. Future agricultural practices may include carbon farming, nutrition-sensitive agriculture, soil-less urban farming, and sky farming. The Sustainable Development Goals of the Agenda 2030 of the UN also depend on protection, restoration and judicious management of soil health.

5.1 Spiritualism in Agriculture and Mystics of Soil among Ancient Cultures

Ever since the beginning of agriculture, humans have looked up to divine powers for advice and help with their agricultural operations such as planting, pest infestation, weather, large harvest, protection of the livestock and the land against natural and human-made calamities. Droughts have been the biggest threat to food security throughout human history and ever since the dawn of settled agriculture about 10–12 millennia ago. Thus, most ancient cultures and tribes believed in deities (gods and goddesses) with divine powers to address specific problems of agriculture. Notable among these were those believed responsible for maintaining soil fertility, good harvest, good seed germination and crop stand, and, of course, abundance of and timely rains to mitigate drought. Examples of such deities worshipped by traditional cultures and tribes are presented in Table 5.1.

Because all terrestrial life depends on soil fertility, Mother Earth is the most widespread concept among traditional cultures across the world. Other deities include those for good harvest, rainfall, soil fertility, grains, agriculture and crops (Table 5.1). Such beliefs are widespread across cultures and those depicted herein are merely some examples.

Major religions of the world have specific prayers for Mother Earth and the need for safeguarding and judicious management of soil and natural resources. Ploughing and soil are prominent parts of several ancient cultures and civilizations. It is thus pertinent that religious organizations emphasize the preaching about soil and its protection in each of the religions. This is an important common point

 Making of a Soil Scientist: From a Subsistence Farm to The Global Forum (R. Lal)
DOI: 10.1079/9781836990154.0005

Table 5.1. Some examples of cultural philosophy that land/soil gifts life. (Adapted from Lal, 2007a, 2013; Evans, 2021)

Deity	Country/culture	God/goddess of:
Atabey (F)	Pre-Colombia America	Mother Earth
Bhu Devi (F)	India	Earth goddess
Ceres (F)	Ancient Rome	Agriculture, crops, fertile land, grains
Demeter (F)	Ancient Greece	Grains, agriculture and fertile soil
Devi Sri (F)	Bali (Indonesia)	Earth, fertility goddess
Dherra (F)	India	Mother Earth
Hequet (F)	Egypt	Flooding of Nile, germination of grain crops; symbolized as a frog
Indra (M)	India	Rain god
Otikan (F)	Turkey	Mother Earth
Pachamama (F)	Andes-Inca	Earth Mother, planting and harvesting
Siya/Sita (F)	India	The furrow
Terra Matre (F)	Italy	Mother Earth
Tudi Gong (M)	China	Harvest

in scriptures of all religions and is the basis of unity among different beliefs and philosophies.

Increase in human population is the major driver of food and nutritional insecurity. Desjardin (2021) estimated that it took 200,000 years for the human population of the world to reach 1 billion (in 1803). It took a further 123 years to reach 2 billion, a further 34 years to reach 3 billion, 15 more years to reach 4 billion and only another 12 years to reach 5 billion. By 2023 it was 8 billion, and future projections (UN, 2024) suggest that it might reach 9 billion by 2037 and stabilize at 10.8 billion by the mid-2080s (Richie and Rhodes-Guiraro, 2024). This rapid expansion caused extinction of several species. Smithsonian National Museum of Natural Historyw (2024) has listed which species went extinct by what year, beginning with 1690. The rate of extinction has increased drastically with rapid increase in human population and the attendant urbanization. Therefore, in the long term, a viable option is to return some agricultural land back to nature (see Chapter 18) through the soil-centric approach being proposed in this book.

5.2 Extinction of Soil

There has been a widespread recognition of the extinction of animals and plants because of the growing and increasingly affluent human population. Rapid disappearance of wild lands has also aggravated the effects of growing interaction between wildlife and humans and the increase in incidence of zoonosis. Risks of zoonotic diseases, infectious diseases that can be passed from animals to humans, are increasing rapidly.

Soil is essential to human existence and the functioning of nature. Soil, as a living and dynamic resource, is the essence of all terrestrial life. Precisely in this context, soil health refers to biological properties in relation to its capacity to provide critical ecosystem services (ESs) for nature and other living beings (i.e. humans). Notable ESs provisioned by soil include clean air, pure water, scenic landscapes, high aboveground and below-ground biodiversity, and high net primary production including crops, forests and livestock, and many more.

Despite its importance as the soil resource for creation of numerous ESs of critical

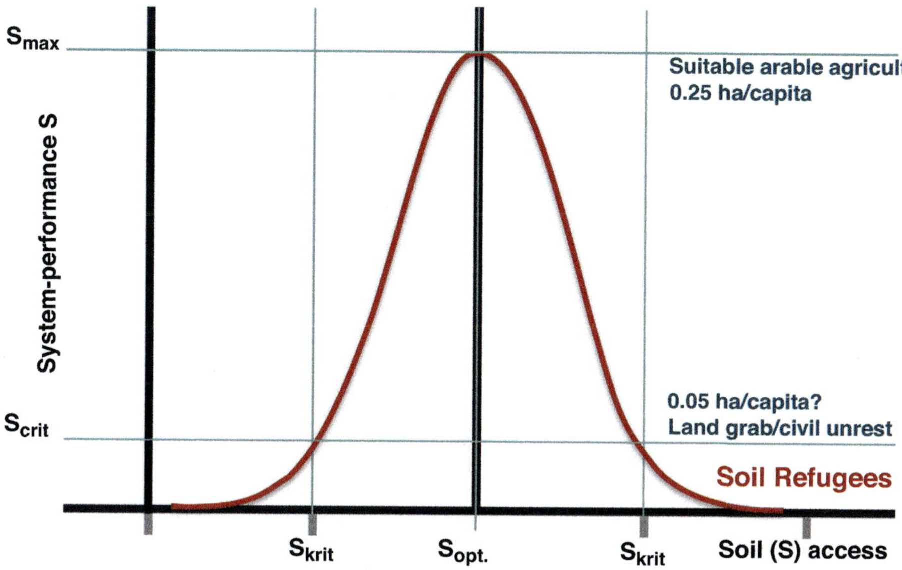

Fig. 5.1. Peak primary soil is a reality in this world with population of 8.2 billion in 2024 and projected to stabilize at 10.8 billion by the 2080s. An adequate amount of per capita agricultural land is estimated at 0.25 ha and its availability in some densely populated countries of South Asia has already reached <0.05 ha. Scarcity of such a prime resource can lead to soil refugees and the problem of land grab, civil unrest, political instability and even war.

importance to humans and nature, there is little if any concern about 'peak' soil, 'endangered' soil or 'extinct' soil. Similar to minerals and other natural resources, soils are also vulnerable to becoming extinct and endangered. Agriculturally, prime soils of the world are finite and fragile, they are unequally distributed geographically at national and global scales and they are highly vulnerable to degradation, pollution and contamination.

The schematic in Fig. 5.1 shows that prime soils may have already passed the 'peak soil' status as measured in terms of per capita land area. For meeting all the basic necessities of life, an optimum level of per capita land area is approx. 0.25 ha. However, the per capita arable land area in some densely populated countries of South Asia may be as low as 0.05 ha. Land scarcity can lead to social and political issues of food insecurity, water insecurity, land grab, soil refugees, political instability, and even war. The downward spiral set in motion by soil degradation can destabilize normal life and lead to political unrest (Fig. 5.2). The sad incidences of land grab and political unrest have drastically increased since the year 2000 and may be aggravated by growing and increasingly affluent human population.

Soils prone to degradation and depletion can reach a peak sooner than those which are protected and have resilience against anthropogenic or natural perturbations. Severity of the problem and widespread distribution can lead to the problems of soil refugees and the attendant socio-economic issues (Fig. 5.2).

The ramification of peak soil is aggravated by socio-economic and political forces. For example, the food insecurity during the 2020s has been aggravated by the interactive and mutually reinforcing nature of other crises that in which the planet is engulfed. These interactive forces are anthropogenic climate change (in conjunction with gaseous emissions by anthropogenic activities); loss of biodiversity (above and below ground); increase in incidence of pests and pathogens; soil degradation; pollution; waste disposal, which aggravates environmental contamination; and, of course, wars and political unrest (Fig. 5.2). Some key interactive factors that aggravate food and nutritional insecurity

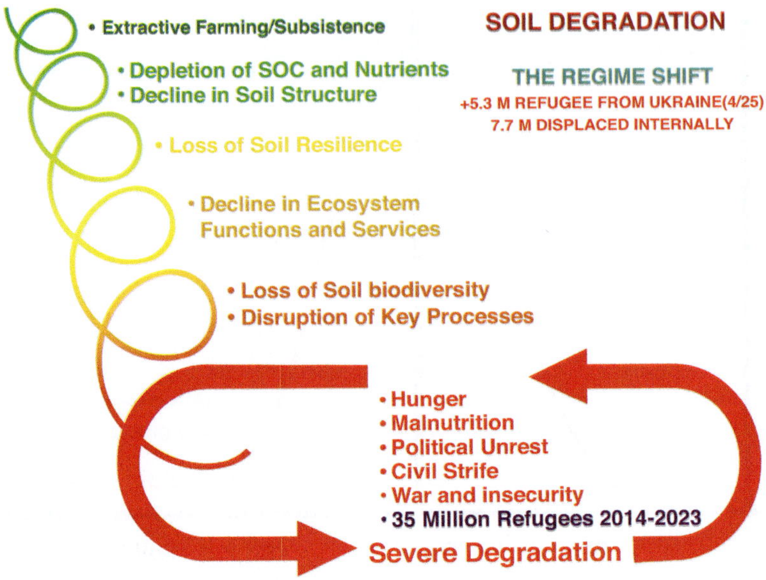

Fig. 5.2. The process of soil degradation set in motion by the depletion of soil organic matter content and the attendant loss of soil structure and other key functions can lead to decline in agronomic productivity and hunger, malnutrition, political unrest, civil strife and refugees migrating to other regions. These problems are aggravated by anthropogenic climate change.

include global climate change, loss of biodiversity, pollution and contamination (degradation) of soil and wars and conflicts (Fig. 5.3). War, like famine, is also a human-made tragedy which aggravates the problem of soil contamination, pollution of water and air, and loss of biodiversity. Soil contaminated by modern war and explosives pollutes the environment and some of these contaminants have a long residence. Food grown on contaminated and polluted soils affects health of all those dependent on those soils.

5.3 Sustainable Development Goals and Policies

The Sustainable Development Goals (SDGs) of the Agenda 2030 of the United Nations (Fig. 5.4) are not on track to be accomplished because soil management is not being given the attention it deserves. This is especially true for SDGs 1, 2, 3, 6, 7, 13 and 15. Hunger and malnutrition are major problems which need to be addressed through protection, restoration and sustainable management of soil resources. In this context, enactment and implementation of the Soil Health Act is necessary at local, state, national, continental, regional and global levels. The aim of the Soil Health Act is to reward farmers for adopting environmentally friendly agriculture. The overall goal is to create policies which are pro-nature, pro-farmer and pro-agriculture, and to transform agriculture from being a problem to part of the solution. The system of rewarding farmers must also be changed, from subsidies (promotion of specific inputs) to payments for ecosystem services (ESs), such as for carbon sequestration in terrestrial ecosystems; improving quality and renewability of water; and strengthening of biodiversity.

Lack of emphasis on soil health and its capacity to generate ecosystem services also had a strong adverse effect on the Development Goals as outlined by Agenda 21 of 1992, the Millenium Development Goals of 2000, and the SDGs of the United Nations Agenda 2030 (Fig. 5.4). Soil health has a strong and direct effect on several SDGs at primary or secondary levels.

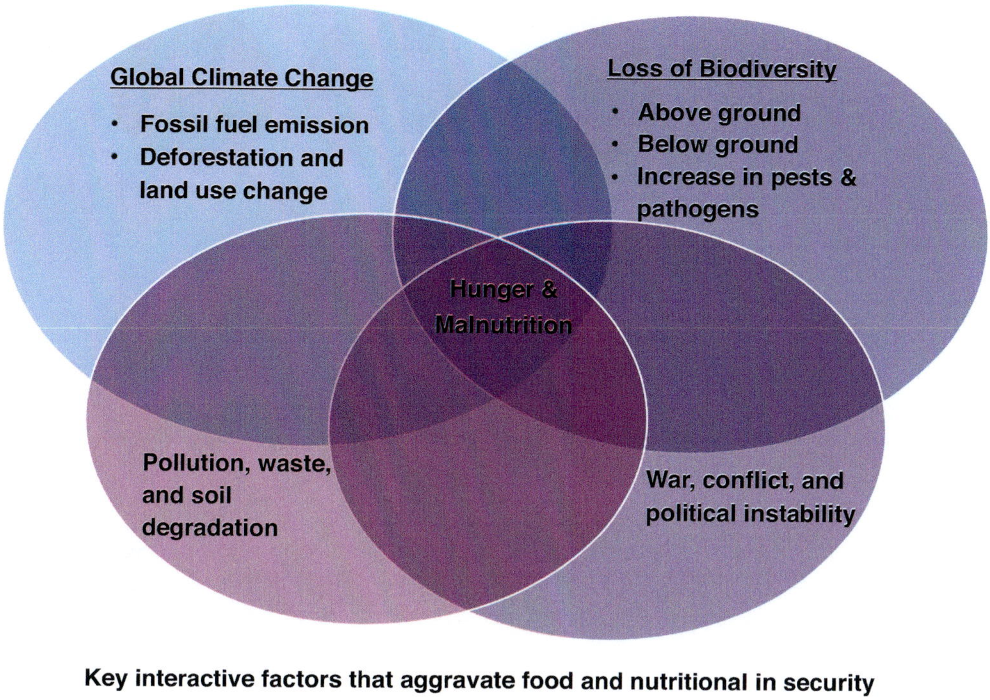

Fig. 5.3. Some key interactive factors that aggravate the problems of undernutrition and malnutrition. Soil degradation and climate change are mutually reinforcing processes and both lead to severity of food insecurity and malnutrition. Restoration of contaminated and polluted soils is a slow process, and it may take generations to denature these pollutants from the soil.

Despite the realization that most SDGs have either direct or indirect effects on the ESs that a soil generates, the primary language of none of the 17 SDGs has a reference to the importance of soil and its functions.

The problem of 'peak soil' when access to prime soil is limited (see Chapter 9) is a widespread issue in developing countries, and Rattan Lal experienced this situation when his family moved from Pakistan to India in 1947. The adverse effects of political unrest on human well-being, aggravated by scarcity of land and water resources, have been the cause of numerous wars throughout human history. These wars over land resources are continuing during the 2020s but the ramifications of war on the planet are not widely recognized nor appreciated.

Any war has three parties: two communities or nations which are at war with one another and the third silent party is the soil on which the war is being fought. Whereas the impacts of war on human miseries and sufferings are widely discussed, the adverse effects of war on nature (wildlife, vegetation, soil properties and functions, water quality and air pollution) are hardly considered. Adverse effects of war on soil and environments can persist for decades or generations and restoration of soil and ecosystems can be a slow process. Any war is a crime against nature and must be avoided.

The adverse impacts of this situation are temporarily masked by the adoption of modern innovations in agriculture (improved varieties, fertilizers and pesticides, irrigation, etc.) but these cannot be eliminated by supplemental inputs. However, long-term management of soil functions and SDGs (as stated in Fig. 5.4) can improve use efficiency of inputs, and enhance production. Protection, restoration and sustainable management of soil is critical to enhancing use efficiency of inputs and enhancing productivity. Soil quality or functions, and more specifically soil health, is the engine of economic

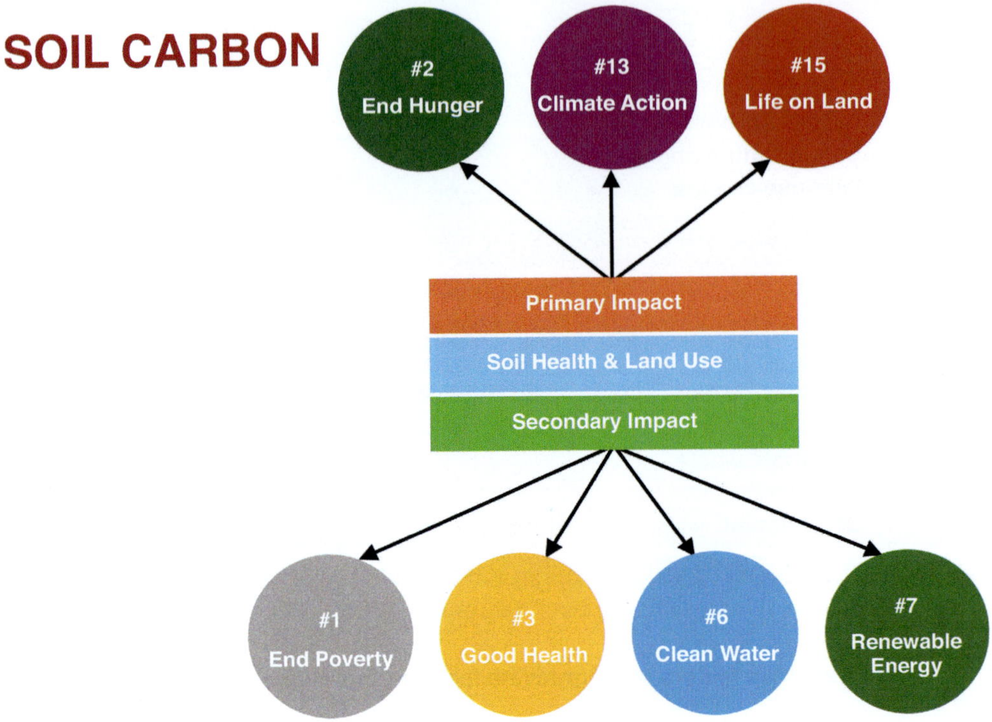

Fig. 5.4. Soil health and the Sustainable Development Goals of the Agenda 2030 of the United Nations. SDGs #2, #13 and #15 are strongly dependent on soil health and its management, whereas SDGs #1, #3, #6 and others have indirect or secondary effects.

development of agrarian economies (Fig. 5.5). Soil health determines the key ESs such as quality and quantity of water; adaptation, mitigation and stabilization of climate change; activity and species composition of above- and below-ground biodiversity; and, above all, quantity and quality of food produced along with its safety and purity. Simply put, global peace is determined by soil health. Those nations which took their soil resources for granted paid a heavy price by their extinction and demise from the face of the Earth. The extinction of historic civilization happened when the population was low and its lifestyle was modest. During the first quarter of the 21st century the population is the driving force in the so-called Anthropocene.

Interest in these issues of the 21st century was created by the conditions experienced by Rattan Lal during his childhood and growing up in a rural community, which was a microcosm of the global issues of the planet during his professional years. By the time he graduated from college in 1963, he realized that there are key soil factors that affect human health and nature's functions (Fig. 5.5). These factors must be protected, restored and sustainably managed for the well-being of all life forms dependent on them. While most farmers, ranchers, foresters and those dependent on them realize the importance of soil, they may neither have the resources nor the access to the essential inputs needed to minimize the risks of soil degradation. The range of physical, chemical, biological and ecological factors interact with one another and are affected by climate change and its uncertainty because of the growing demands of humanity. Humans are the primary drivers of these factors, and thus the prominent determinant of the so-called Anthropocene that may have begun with the onset of settled agriculture rather than with that of the Industrial Revolution (Crutzen, 2010).

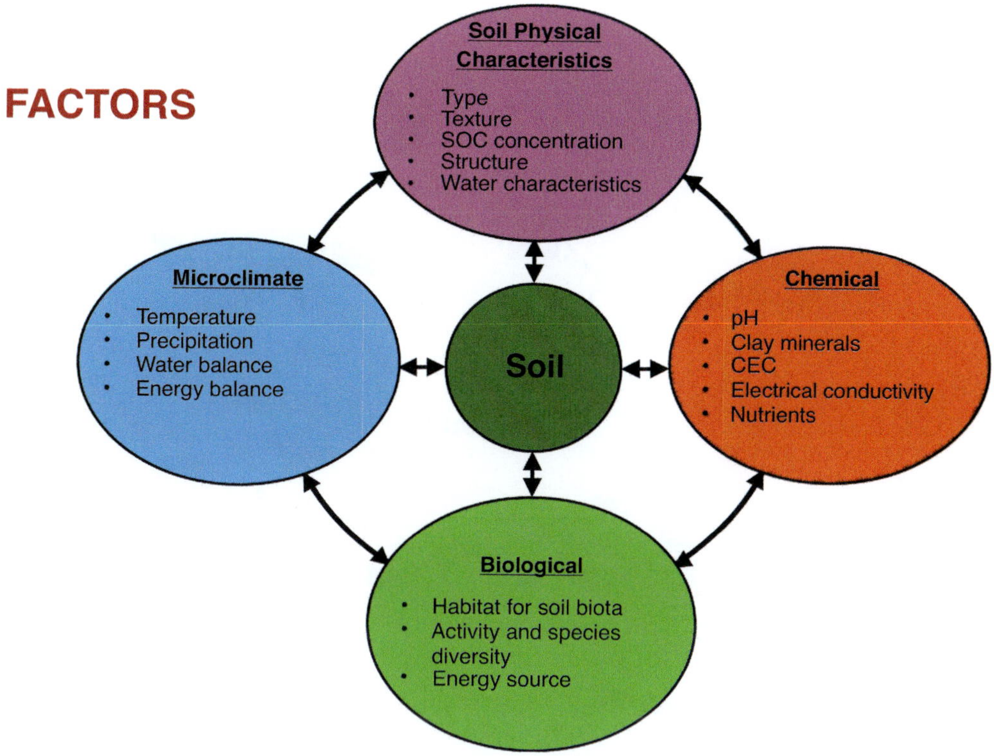

Fig. 5.5. Soil and its attributes which affect use efficiency of inputs and the overall functions and productivity of a specific land use or farming system.

Key factors of soil degradation outlined in Fig. 5.5 are also related to soil health. The latter being a biological property, dependent on activity and species diversity of soil biota, is strongly affected by soil organic matter (SOM) content and its dynamic as affected by land use and management (Fig. 5.6). Indeed, soil health is the engine of economic development because of its role in creating and strengthening some key ecosystem services.

Some major ESs determined by SOM content and its dynamics are water quality and quantity, climate change adaptation and mitigation, food and nutritional security and safety, and biodiversity above and below ground. Thus, the finite and fragile soil resources must never be taken for granted. While SOM is the engine of economic development, its relative proportion in the solid fraction is small and easily depleted (Fig. 5.7).

5.4 Modern Gods and Goddesses of Soil and Agriculture

The ancient gods and goddesses have been replaced by some key disciplinary terms used in modern scientific literature. These are: pedology, agronomy, agroecology, biotechnology, plant physiology, climatology, etc. Indeed, adoption of scientific innovations has alleviated some of the constraints faced by farmers. Thus, agronomic productivity has drastically improved since the middle of the 20th century. Indeed, agriculture has a bright future ahead. More changes will occur between 2020 and 2050 than have occurred since the beginning of agriculture over the past 12 millennia through building scientific innovations on top of traditional knowledge and wisdom. These scientific developments can alleviate the soil-related constraints to achieving high agronomic yield.

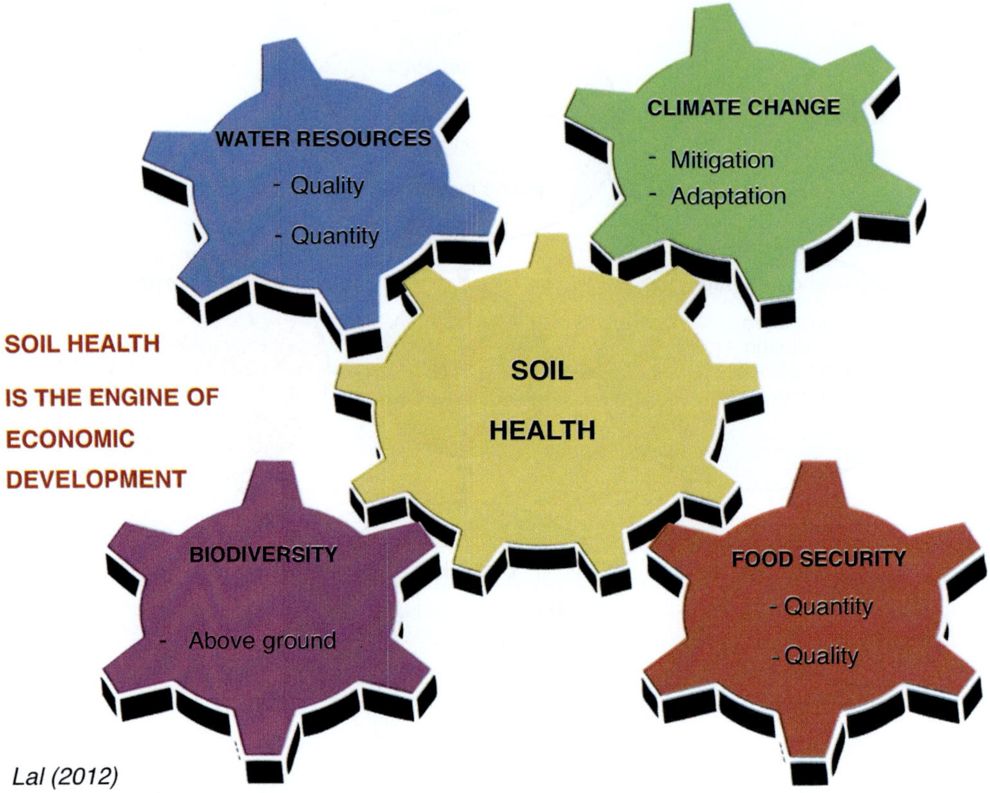

Fig. 5.6. Soil health is the engine of economic development of an agricultural economy. (Adapted from Lal, 2012)

5.5 The Future of Agriculture

The historic past and foreseen future of agriculture are shown in a schematic in Fig. 5.8. Agriculture experienced four revolutions: revolution 1 was the transition from hunter-gatherer to settled agriculture; revolution 2 was the development of hand tools around 1750 and mechanized farming around the 1930s; revolution 3 was the Green Revolution of the 1960s based on the use of input-responsive varieties, and of the 2000s on the adoption of a soil-centric approach to sustainable intensification; and revolution 4 is the use of artificial intelligence and robotics since about 2010, involving digital agriculture and precision farming. Other components of revolution 4 are sky farming, nutrition-sensitive agriculture, carbon farming and soil-less urban food production systems.

5.6 Soil Management Beyond Food and Fuel

Soil is a major reservoir of terrestrial carbon stock and a source of green water in the root zone (Fig. 5.7). While carbon concentration is limited in the surface – 1 m depth for SOC and 0.5–2 m depth for SIC – the entire soil profile (surface and subsoil) is the store of the so-called green water on which depends all terrestrial life. All ESs essential for nature and humans depend on the entire soil profile, especially those which are useful for all terrestrial life (macro-, meso- and micro-flora and -fauna).

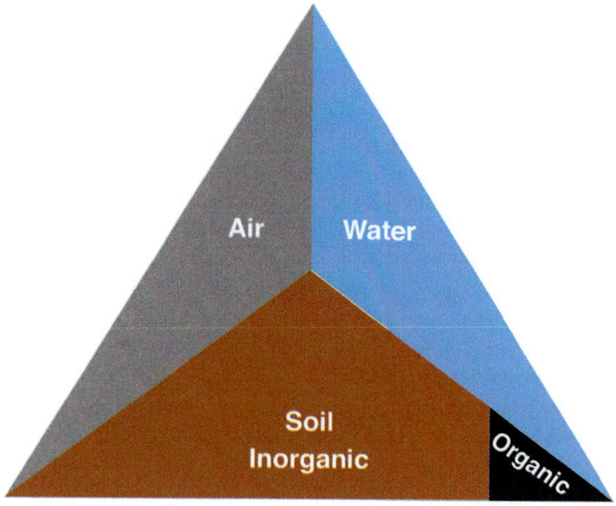

Fig. 5.7. The small magnitude of the organic fraction in the solid component of the soil plays a critical role in generating and sustaining some key ecosystem services essential to human well-being and nature conservancy.

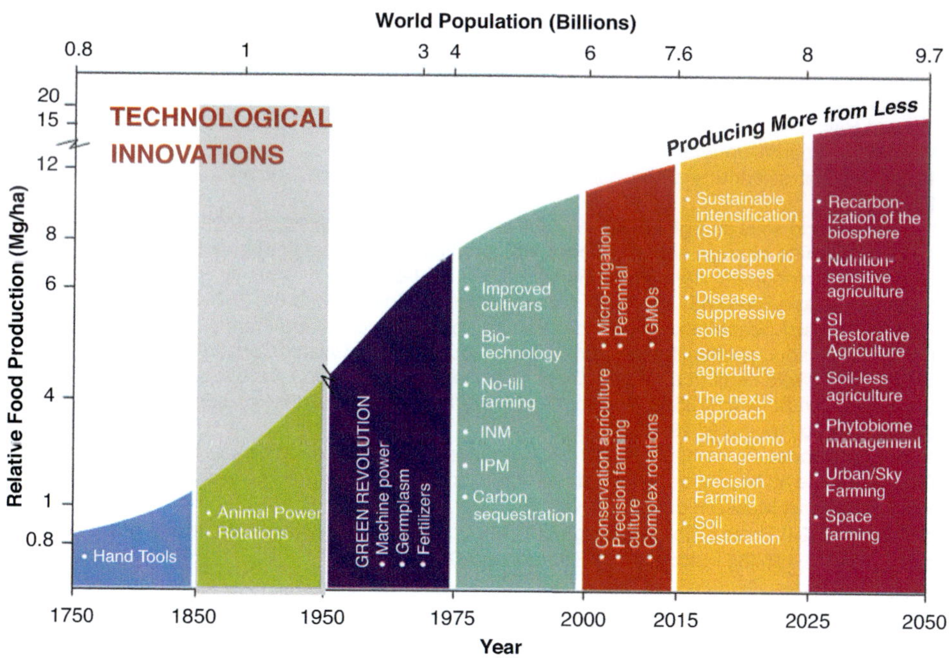

Fig. 5.8. Technological innovations of the historic past, which supported humanity, and future innovations projected to happen between 2025 and 2050 that will radically transform agriculture from being a problem to being a part of the solution to addressing environmental and related issues of the 21st century.

Thus, soil is the source of numerous ESs beyond agriculture or 4 Fs involving production of food, fuel, fibre and feed. Other critical ESs include: (i) sequestration of atmospheric CO_2 in terrestrial sinks (soil and vegetation) for adaptation and mitigation of climate change; (ii) improving quality and renewability of water; (iii) strengthening of biodiversity through improvements in activity and species diversity of soil biota; (iv) harnessing of raw materials for industries other than agri-based systems; (v) production of antibiotics and other medicines for humans and livestock; (vi) use of soil for aesthetics, art and culture; and (vii) studies of human and planetary history, among others. In future, study of soil science will focus on these ecosystem services which are not listed in the Fig. 5.8.

5.7 Soil and Spirituality

Rattan Lal used a broad-based and holistic approach to sustainable management of soil resources. He combined religious philosophy with the scientific advances to highlight the importance of soil, not just for humans but for all terrestrial life. This is where science of soil meets the philosophy of all major religions of the world. Indeed, there is a strong link between soil, soul, spirituality and stewardship (Lal, 2024). Indeed, the problems facing humanity and the planet are so enormous that there is no one simple solution. Indeed, global peace and political stability depend on soil health and soil management.

Bibliography

Crutzen, P.J. (2010 Anthropocene man. *Nature*, 467(7317), S10–S10. https://doi.org/10.1038/467S10a

Crutzen, P. and Stormer, E.F. (2000) The Anthropocene. *Global Change Newsletter* 41, 17–18.

Desjardins, J. (2021) Population Boom: Charting how we got to nearly 8 billion people. https://www.weforum.org/stories/2021/12/world-population-history/

Evans, M. (2021) *Soils: The Incredible Story of What Keeps the Earth and Us Healthy*. Murdoch Books, Sydney, Australia. ISBN: 9781922351418

Handelsman, J. (2021) *A World Without Soils: The Past, Present and Precarious Future of the Earth Beneath Our Feet*. Yale University Press, New Haven, Connecticut. ISBN: 9780300256406

Jenik, J. and Hall, J.B. (1966) The ecological effects of the Harmattan wind in the Djebobo Massif (Togo Mountains, Ghana). *Journal of Ecology* 54(3), 767–779.

Lal, R. (2007a) Soil science and the carbon civilization. *Soil Science Society of America Journal* 71, 1095–1108.

Lal, R. (2007b) World soils and global issues. *Soil and Tillage Research* 93, 237–238.

Lal, R. (2012) Climate change and soil degradation mitigation by sustainable management of soil and other natural resources. *Agricultural Research* 1(3), 199–212.

Lal, R. (2013) Soil and Sanskriti. *Journal of Indian Society of Soil Science* 61, 267–274.

Lal, R. (2015) The soil-peace nexus: Our common future. *Soil Science and Plant Nutrition* 61, 566–578.

Lal, R. (2024) Soil, soul, spirituality and stewardship. *Journal of Soil Water Conservation* 79(1), 10A.

Quicquid plantatur solo cedit (n.d.) *A Law Dictionary: Adapted to the Constitution and Laws of the United States by John Bouvier* (1856). Available at: https://legal-dictionary.thefreedictionary.com/Quicquid+plantatur+solo (accessed 29 May 2025).

Richie, H. and Rhodes-Guiraro, L. (2024) Peak global population and other key findings. Our World in Data. Available at: https://ourworldindata.org/un-population-2024-revision (accessed 29 May 2025).

Smithsonian National Museum of Natural History (2024). Washington, DC.

United Nations (2024) Three billion people globally impacted by land degradation. https://news.un.org/en/story/2024/12/1157651 (accessed 2 September 2025).

6 The Importance of Mentoring

'A mentor is someone who sees more talent and ability within you than you see in yourself and helps bring it out of you.'
— Bob Proctor

'A mentor is someone who allows you to see the hope inside yourself.'
— Oprah Winfrey, in an interview on WCVB-TV 5 News CityLine (Boston, 13 January 2002)

'Puoi solo aiutarlo a scoprire ciò che ha dentro di sé.'
('You cannot teach a man anything. You can only help him discover it within himself.')
— Galileo Galilei

Abstract

Mentoring is important to professional development. Rattan Lal was privileged and blessed in receiving mentoring from dedicated professionals who were also role models because of their own accomplishments. In this context, the importance of mentoring during the early stages of education (primary, middle and high school) can never be over-emphasized. Mentoring during the early education is the foundation of moral, ethical, academic and professional development. Subsequent college and postgraduate mentoring develops on the foundation of the school level, and it is carried on during the initial stages of a professional career. During his professional career as a soil scientist, Rattan Lal had a great regard for his college professors who taught him soil science. His professional goal since 1987 was to transform agriculture from being a problem into a part of the solution to advance global security for food, water, climate and other essential ecosystem services. This goal was made possible by his exposure to the philosophy and accomplishments of some historic scientists, contemporary scientists and pioneering authors of environmental movements.

6.1 Mentoring During Primary and Secondary School

Mentoring, at all levels, from primary school to graduate education and beyond, during the early stages of one's career, is critical to professional development, and its importance can never be over-emphasized. Rattan Lal had the good fortune to have excellent, dedicated and caring teachers at all levels of education and during his early career. During primary school, he had a fourth-grade teacher who played an important role not only in enhancing his interest in mathematics (algebra) and other subjects, but also in teaching him yoga and the related exercises, including the *pranayama* (breathing in and holding the breath), but more importantly, some mantras for meditation. Rattan Lal has been a strong believer in the recitation of the Gayatri mantra, which helped him during times of anxiety and fear of both the known and the unknown. He learned how to speak in front of a class or school gathering from his Sanskrit teacher in seventh and eighth grades. The teacher was an excellent speaker and quoted many verses from scriptures. Rattan Lal followed the style of his teachers in public speaking and expressing his views while quoting writings of historical personalities (Budha, Gandhi), ancient scriptures (e.g. Vedas, Puranas, *Gita*, *Ramayana*, *Mahabharata*, *Upanishads*, Quran, Gurbani, Bible) and critical

 Making of a Soil Scientist: From a Subsistence Farm to The Global Forum (R. Lal)
DOI: 10.1079/9781836990154.0006

thinkers (e.g. Jawaharlal Nehru), and other political and educational leaders at domestic, national and international levels.

6.2 Mentoring at the College Level

6.2.1 Student at the Government Agriculture College and Research Institute, Ludhiana (1959–1963)

Rattan Lal had the good fortune to be admitted into the BSc (Agric.) programme at the Government Agricultural College and Research Institute, Ludhiana, in the class that graduated in 1963. Being a student provided him not only with an opportunity to study but also to learn the global importance of agriculture as a profession. He also had a great opportunity to develop his skills as a speaker, and took an active part in the Speaker's Union for activities such as debate and declamation contests, extempore speaking, and poetry recitation. These were the only extracurricular activities that he participated in during his undergraduate education.

During the very first year of his college, at age 15, Rattan Lal was highly motivated to study hard and was determined and focused on his goals. His motivation was based on the goal of earning an academic scholarship given to the top five students of the class along with receiving the refugee grant. The desire to win these two scholarships made him competitive in scholarship and in the habit of working hard to achieve his goals. He was very focused on these goals and had no other mission.

Professors who motivated him during undergraduate studies were Prof. R.S. Vasudeva and Prof. H.R. Kalia in botany, Prof. B.N. Sharma and Prof. S.L. Chopra in chemistry, and Prof. D.R. Bhumbla and Prof. J.S. Kanwar in soil science. Dr Bhumbla, a recent returnee from the Ohio State University (1961–1962) was a great motivator and helped Rattan to join the MSc degree programme at the Indian Agricultural Research Institute, New Delhi, and then to seek admission for a PhD degree at the Ohio State University.

During his final-year examination, Rattan Lal received the highest grade point in his class and his name was written on the Roll of Honour in the College at Ludhiana for the year 1963 (Fig. 2.4). Achieving the first position was an important motivator for Rattan Lal during his future education and in his professional career; it gave him the confidence to compete. By the time of his graduation, the college was upgraded to the Punjab Agricultural University (PAU) and Rattan Lal's class was the first class to receive a diploma from PAU in its first convocation in 1963 (Fig. 2.3).

6.2.2 Postgraduate School at the Indian Agricultural Research Institute (IARI), New Delhi (1963–1965)

With the encouragement and guidance of Dr D.R. Bhumbla, Rattan Lal applied for admission into the MSc programme in the Division of Soil Science and Agricultural Chemistry at IARI for the class to graduate in 1965. He was one of the five students admitted. His four class fellows included those who also received first position in their respective degrees from their institutions in Uttar Pradesh, Rajasthan etc., and one of the students was from egypt. There were also students from other countries such as Nepal, Thailand and the USA. IARI had faculty and students from throughout India. It was the first time Rattan Lal had met students and faculty who spoke other languages of India (e.g. Tamil, Bengali, Marathi, Telgu, Malyalam), and also the first time that he ate food in a college canteen. The academic competition was intense, and at stake were two gold medals presented to the student who earned the highest grade point average. In addition to the medals, there was also the question of being admitted for the PhD degree and also earning a fellowship for pursuing the degree at IARI. Being at IARI also provided an opportunity to meet scholars who visited IARI from abroad. This opportunity enabled Rattan Lal to broaden his outlook on national and international issues.

Rattan Lal obtained the highest grade point and thus earned two gold medals from IARI (Fig. 2.5). By now, he had earned the highest grade point in his high school (matriculation), and in his BSc and MSc degrees. This confidence to be able to compete proved to an important factor in his professional career.

6.2.3 PhD student at the Ohio State University (1965–1968)

Rattan Lal arrived in Columbus, Ohio, on 29 December 1965 with US$8 (obtained at New Delhi Airport for exchanging Rs 40) in his pocket and hopes of earning a PhD degree from his dream university, the Ohio State University (OSU). The classes began on 3 January 1966. In addition to the bitter winter, Rattan Lal faced the challenges of language, food habits and culture shock. Everything was different. The Office of International Affairs of Ohio State was very helpful in arranging a loan and an apartment as a paying guest with an American host family. After about 1 week's stay with students from PAU who were studying at Ohio State under the PL-480 scheme, Rattan Lal moved in with his host family. The elderly couple were very kind and helpful. They even invited him to attend church with them, but he did not go because studying and completing the homework were his highest priority.

Rattan Lal had to take 1 year of classes in English (both written and spoken). His pronunciation and manner of speaking, a combination of Hinglish and Pinglish, was a major challenge. Any time he went to the main office of the Department of Agronomy, the staff always gave him a paper and a pencil to write down what he wanted. On one such visit, a staff member looked at his writing and said quietly, 'My goodness, I can only read what is written in English alphabets!'.

Rattan Lal started his field research and established a drainage experiment at the Waterman Farm at a location which is now under the Martha Morehouse facility on the West Campus of The Ohio State University. Rattan often walked from Townsend Hall (1885 Neil Avenue), which housed the Department of Agronomy, to the field site. Prof. G.S. Taylor, his son Ron and the farm manager (Lloyd Roof) also provided transport and helped in the logistics and fieldwork.

One notable incident was when the young seedlings of maize, four per lysimeter (Figs 2.6–2.8), were killed perhaps by over-fertilization in a lysimeter or with some herbicides in the fertilizer. Dr Taylor was very disappointed that one growing season was lost. Rattan Lal decided to transplant seedlings from the surrounding field and looked after these for a week until they took root and were growing satisfactorily. About 2 weeks after transplanting, Rattan Lal imposed the drainage treatments of maintaining the water table at 15 cm and 30 cm depths and fully drained. Within 2 weeks of imposing the treatments, Rattan Lal could see the differences in leaf colour, plant height, adventitious roots and scalding (wilting because of low O_2 content in soil air) in the afternoon for the poorly drained treatments. He also monitored the composition of soil air. It was then that he invited Prof. Taylor to come and see the experiment, and the professor was not sure what to expect. Success of the experiment, corn growing in the lysimeters and exhibiting distinct symptoms of anaerobiosis for the shallow water table treatments, was the biggest surprise to Prof. Taylor. From now on, Rattan Lal was one of the favourite students in the department. Perseverance, hard work and total commitment to the task at hand, paid the highest dividends. Rattan Lal defended his PhD thesis in August 1968, just prior to his official 24th birthday.

This research on the effects of water table depth on crop growth and the soil processes under anaerobiosis (poor drainage and lack of oxygen in the root zone) turned out to be an important lesson for Rattan Lal in his later (early 1990s) professional interests in gaseous emissions from soil (e.g. CH_4 and N_2O with the global warming potential of 28 and 310, respectively, compared with that of CO_2) as a cause of anthropogenic global warming (Lal and Taylor, 1969). As a faculty member since July 1987 at OSU, Rattan Lal developed on this PhD thesis experience and promoted the idea of sustainable management of soils so that agro-ecosystems are part of the solution rather than a problem of environmental degradation in general and mitigation and adaptation of anthropogenic climate change in particular. This was the foundation of his future research on soil and climate with the focus on making soil a sink of atmospheric CO_2 and using improved agro-ecosystems as a strategy for adaptation to and mitigation of climate change (Lal, 2004). His professional goal since 1987 has been to transform agriculture from being a problem into a part of the solution to advance global security for food, water, climate and other essential ESs.

6.2.4 Senior Fellow at the University of Sydney, NSW, Australia

This section is a follow-up of the section 'High and Low Tea in Sydney' (Chapter 2). Rattan

Lal had a unique opportunity to learn different concepts and philosophy about soil science by working in the Department of Soil Science at the University of Sydney, which followed from what he had learnt at the Midwestern Land Grant University system in Ohio. Yes, Prof. Baver and Prof. Taylor were right. Rattan Lal studied thermodynamics of soil moisture retention and movement in vertisols (clayey soils with high swell/shrink capacity and water retention but slow water infiltration and hydraulic conductivity) and found that the heat of wetting released during water infiltration into dry soil was the cause of slaking, low infiltration rate and risks of accelerated water run-off and hydric soil erosion (Collis-George and Lal, 1971, 1972). Thus, the logical solution to this serious problem affecting water infiltration in vertisols was retention of crop residues on the soil surface as mulch in combination with no-till farming or conservation agriculture. Rattan Lal learned the importance of understanding the basic processes of soil structure and water movement in swelling soils and how to manage them.

During his stay in Sydney, Rattan Lal was also fortunate to meet Prof. Dennis Greenland (Fig. 6.1). He was a visiting professor on sabbatical leave from the University of Reading, UK, and was based at the Waite Research Institute in Adelaide, Western Australia. He and Prof. T.J. Marshall came to Sydney to review the research programme of Rattan in 1969, and this visit established a long-term co-operation with Prof. Greenland. As luck would have it, Rattan Lal accepted the position of soil physicist at IITA in Ibadan, Nigeria. He arrived at IITA in December 1969, and Prof. Greenland returned to the University of Reading and was appointed a member of the Board of Trustees of IITA. This was a fortunate professional coincidence for Rattan who now had an opportunity to work under the guidance of Prof. Greenland.

6.2.5 Soil Physicist at IITA

The offer of the position of Soil Physicist at IITA was made to Rattan Lal by the first Director General of IITA, Dr Herb R. Albrecht. It so happened that Dr Albrecht had interviewed Rattan Lal in 1968 in the office of Dr Garth Volk, then Head of the Department of Agronomy at the Ohio State University. While Rattan Lal moved from Sydney to IITA, Prof. Dennis Greenland had returned from Adelaide to Reading and was appointed a member of the Board of Trustees of IITA, and attended the first meeting of the board in April 1970. This was where he met Rattan Lal who was now studying soil physical properties under the tropical rainforest and under shifting cultivation. As Prof. Greenland noted, Rattan Lal had made a radical transformation from someone studying thermodynamics of soil moisture retention and movement in the clayey soils of NSW to practical and on-field assessment of physical properties of soils of the West African humid tropics without the support of any laboratory facilities. The best outcome for Rattan Lal was that he worked at IITA with Prof. Greenland for about ten years from 1970 to 1980. In 1975, Prof. Greenland was appointed Director of Research at IITA, and he mentored Rattan Lal to study processes of soil erosion and adopt no-till farming as an effective soil and water conservation measure for structurally fragile soils of West Africa (Greenland, 1975; Lal, 1987). Working at IITA from December 1969 to July 1987 was a great research experience for Rattan Lal. His 18 years of research at IITA can be summarized as follows:

- Do not disturb soil surface.
- Keep soil always covered with crop residue mulch or a cover crop.
- Grow deep-rooted legumes (pigeon peas, stylosanthes, mucuna) in the rotation cycle.
- Grow food crops (cereals, legumes, root crops) by conservation agriculture.
- Integrate crops with trees and livestock.

Now Rattan Lal was an independent scientist with experience of working in India, USA, Australia and Africa. From being a mentee, Rattan was now a mentor of young soil scientists with the overall focus of sustainable management and agriculture for soils of the humid, sub-humid and semi-arid tropics.

In his professional career as a soil scientist, Rattan had a great regard for his college professors who taught him soil science. Notable among these were Dr D.R. Bhumbla and Dr J.S. Kanwar, both from the Punjab Agricultural University, Ludhiana; Dr B. Ramamurthy and Dr T.D. Biswas, from the Indian Agricultural Research

Fig. 6.1. Prof. D.J. Greenland, a globally eminent soil scientist of his era, was Head of the Department of Soil Science at the University of Reading in U.K., and Director of Research at IITA (1975–1977). (Photo: Mary Greenland and Dr Alfred Hartemink)

Institute (IARI), New Delhi; Dr George S. Taylor and Dr L.D. Baver, from the Ohio State University; Dr D.J. Greenland from the University of Reading and IITA; and Prof. N. Collis-George from the University of Sydney, Australia. The latter turned out to be very influential in innovative thinking of scientific issues and in technical writing. Dr G.S. Taylor was his PhD mentor, a well-wisher and responsible for Rattan's return to the Ohio State University as a faculty member. He made it possible for Rattan to be appointed to the position held by Prof. George S. Taylor. He was the professional and academic 'father' of Rattan Lal. He helped the Lal family get settled on their arrival at Ohio State in 1987 and was like a grandfather to Rattan's children. Prof. Taylor's role in Rattan's faculty career at The Ohio State University can never be over-emphasized.

6.2.6 Mentoring at the professional level

At the college and postgraduate levels, Rattan was exposed to the philosophy and accomplishments of some historic scientists such as: Sir Albert Howard (1943), contemporary scientists (e.g. Dr Norman Borlaug, Dr M.S. Swaminathan (2006), Dr Gurdev Khush, Dr Raj Paroda, and pioneering authors of environmental movements (e.g. Pearl Buck, 1931; John Steinbeck, 1939; Rachel Carson, 1962; Barry Commoner, 1971).

Bibliography

Buck, P. (1931) *The Good Earth*. John Day, New York and Methuen, UK.

Carson, R. (1962) *Silent Spring*. Houghton Mifflin, Boston, Massachusetts. ISBN: 0395683297

Chatterjee, R.K. and Lal, R. (1967) Growth and yield of maize as influenced by soil physical properties. *Indian Journal of Agricultural Science* 38, 816–819.

Collis-George, N. and Lal, R. (1971) Infiltration and structural changes as influenced by initial moisture content. *Australian Journal of Soil Research* 9, 107–116.

Collis-George, N. and Lal, R. (1972) The temperature profile of soil column during infiltration. *Australian Journal of Soil Research* 11, 93–105.

Commoner, B. (1971) *The Closing Circle: Nature, Man and Technology*. Dover Publications, Inc, Mineola, New York.

Gandhi, M.K. (1948) *The Story of My Experiment with Truth*. Public Affairs Press, Washington, DC. ISBN: 817229008X

Greenland, D.J. (1975) Bringing the Green Revolution to the shifting cultivators: Better seed, fertilizers, zero or minimum tillage, and mixed cropping are necessary. *Science* 190, 841–844.

Howard, Sir Albert (1943) *An Agricultural Testament*. Oxford University Press, London.

Lal, R. (1987) Managing soils of sub-Saharan Africa. *Science* 236, 1069–1076.

Lal, R. (2004) Soil carbon sequestration impacts on global climate change and food security. *Science* 304, 1623–1627.

Lal, R. and Taylor, G.S. (1969) Drainage and nutrient effects in a field lysimeter study. I. Corn yield and soil conditions. *Soil Science Society of America Proceedings* 33, 937–941.

Nehru, J.L. (1946) *The Discovery of India*. Penguin Random House, India.

Steinbeck, J. (1939) *The Grapes of Wrath*. Penguin.

Swaminathan, M.S. (2006) An evergreen revolution. *Crop Science* 46(5), 2293–2303.

Winfrey, O. (2002) Radio broadcast, WCVB-TV 5 News CityLine. 13 January.

7 Potential and Opportunities for Professional Advancement

'We cannot predict the future, but we can invent it.'
(Original: 'The future cannot be predicted, but futures can be invented.')
— Orville Freeman, U.S. Secretary of Agriculture quoting Dennis Gabor

'Quando io crederò imparare a vivere, e imparerò a morire.'
('While I thought I have been learning how to live, I have been learning how to die')
— Leonardo Davinci, Codex Atlanticus *(Interpretation: 'Learning never exhausts the mind.')*

Abstract

In addition to advice from world-class mentors, Rattan Lal was fortunate in having a hidden motivator – hard work and commitment to excellence. The source of this motivation was the resource-poor small landholder farmers who were vulnerable to risks of food insecurity because of harsh climate, degraded and depleted soils, and lack of resources needed to enhance agronomic yield. He was fortunate in having working experiences at several universities and with the CGIAR system. This unique combination of working under diverse academic and eco-regional environments was extremely useful. Rattan Lal focused on his scientific mission of research, education and extension, and on publishing his data in journals around the world. His audiences were students, researchers, extension agents and practitioners. He believed that science cuts across religious, cultural, ethnic and political boundaries.

Rattan Lal received many prestigious international awards. He believed that these awards were recognition of the soil science profession and not of himself, his argument being that he was merely a representative of 65,000 soil scientists. This was the reason that he donated the honoraria from all awards to the Ohio State University for strengthening research, education and outreach in soil science, and support international Scholar Exchange.

7.1 Benefits of Working with the CGIAR System

The Consultative Group in International Agricultural Research (CGIAR) was launched in 1974 (Lele, 2021; Gates, 2023). Rattan Lal was fortunate and privileged to be appointed as a scientist in the CGIAR system, specifically as a soil physicist at the International Institute of Tropical Agriculture (IITA), Ibadan, Nigeria. He arrived in Ibadan in December 1969, about four months prior to the official inauguration of the centre. It turned out to be a unique opportunity to address challenges faced by small landholder and resource-poor farmers who had been practising shifting cultivation and bush fallow rotation while cultivating some fragile soils under harsh tropical climates of sub-Saharan Africa (SSA) and elsewhere in the humid, sub-humid and semi-arid regions of the tropics and sub-tropics (Nye and Greenland, 1965).

By 1970, there were a total of four international agricultural research centres: International Rice Research Institute (IRRI), established in 1960 at Los Banos, Philippines; International Maize and Wheat Improvement Center (CIMMYT), established in 1963 at El Batan, Mexico; International Center for Tropical Agriculture (CIAT), established in 1967 at Los Palmas, Colombia; and International Institute of Tropical Agriculture (IITA), established in 1970 at Ibadan, Nigeria. The CGIAR was created in 1974 to manage these

DOI: 10.1079/9781836990154.0007

four centres. It comprised the Ford and Rockefeller Foundations, along with the World Bank, the FAO and the UNDP. By 2000, there was a total of 18 centres within the CGIAR system. In 2019, the CGIAR system was reorganized into one CGIAR system with management under a central organization (Lele, 2021). It is important to recognize that the CGIAR system had been successful in addressing the global issue of food insecurity, especially in developing countries.

By joining IITA in 1969, Rattan Lal was a professional soil scientist, and being appointed at IITA created an opportunity for him to secure his place among the global scientific community. To Rattan Lal, who was motivated in the decade of his college/university education (1959–1968) to compete for scholarship, the challenge was to survive on his own and compete for his place among professionals in his discipline of soil science. The CGIAR system, in addition to addressing the issues of food security, also created some world-class researchers. Notable among these, who also won the prestigious World Food Prize, were Dr M.S. Swaminathan, Dr S. Raja Ram and Dr Gurdev Khush. There were also others who achieved administrative positions and who were very successful in their mission to become administrators, such as Dr Nyle Brady. Working at IITA for about 18 years provided Rattan Lal with a unique opportunity to conduct field-scale research on important thematic topics of: soil degradation by accelerated erosion; effects of methods of deforestation and land clearing on micro- and meso-climate; and impact of mulch farming and cover cropping on soil productivity and sustainability. During his tenure at IITA, Rattan conducted research on the importance of soil physical properties and processes on crop growth and the environment. Notable among these properties were soil temperature, drought stress, soil compaction and crusting, and adverse effects of ploughing on soil quality. He also conducted research on no-till farming, rates of mulching on soil erosion control, and soil macrofauna (e.g. earthworms). He conducted experiments on watershed management, and advanced scientific knowledge on sustainable management of soil and other natural resources. He published his research in international journals (Lal, 1987) and organized several international conferences on scientific issues of the time (e.g. soil erosion in the tropics (1975); sustainable management of soil physical properties (1977); tropical deforestation in relation to alterations of elemental and hydrological cycling along with changes in energy balance (1981)).

While at IITA and working within the CGIAR system, Rattan Lal had a unique opportunity to meet and work with world-class soil scientists of the 1970s and 1980s. Notable among these were Dr Danny Hillel, 2010 World Food Prize Laureate (Israel); Dr Wilfred Gardener, Dr Nyle Brady, Dr Bill Moldenhauer, Dr Pedro Sanchez, and Dr Samir El-Swaify (USA); Prof Norman Hudson and Prof Walter Russell (UK); Dr Ildefonso Pla Sentis (Venezuela); Dr Eric Roose (France), and many more. While working at IITA, Rattan Lal laid the foundation of some key concepts in sustainable agriculture in the tropics: (i) using a soil-centric approach to green revolution; (ii) producing more from less through eco-intensification; and (iii) transforming agro-ecosystems into nature-positive and negative emission technologies (NET). Staying focused on research paid numerous hidden benefits to Rattan Lal that neither he nor anyone else could have realized otherwise. Indeed, dedication to science and commitment to excellence are the only paths to accomplish the mission of serving humanity while protecting Mother Earth (*Terra Matre*).

The eighteen years of work at IITA exposed Rattan Lal to the global forum for addressing issues related to food and the environment. It was this experience that motivated him to study anthropogenic climate change and how agro-ecosystems can be used as a sink for atmospheric CO_2. Indeed, working at IITA and knowing the research of the CGIAR system was the foundation of the new concept of 'carbon farming' on which he focused during his tenure at the Ohio State University, managing a large project funded by FFAR from 2022 to 2027.

7.2 Opportunities Provided by the Land Grant University System

Ten years of university education in three continents (6 in South Asia, 3 in the USA, 1 in Australia) and then 18 years with the CGIAR created unique opportunities for Rattan Lal to develop a programme of research and education at a global level. The experience gained at IITA was the foundation for developing a successful teaching and research programme at The Ohio State University. Over the 40-year period at OSU, Rattan Lal worked with approx. 425 researchers from almost 60 countries. It was these scientists from around the world who created the global programme on soil management and transformed Rattan Lal from someone raised on a subsistence farm to one who spoke at global forums on soil and environmental issues. He believed the credit for this successful transformation was due to these 425 researchers. The combined experience of working with the CGIAR system and the Land Grant University system was the unique feature of his career and the basis of his success.

7.3 Getting the U.S. Visa

With four young children of school age, Rattan Lal had a very pressing need to find a job that would provide education for them and create an opportunity for him to accomplish his three scientific missions stated above. Prof George S. Taylor, his PhD advisor at OSU, who retired in 1984, encouraged him to apply for the position vacated by him at the Department of Agronomy at OSU. Rattan Lal applied for the position and visited OSU for the interview during the fall of 1984. At the time, there were two vacancies in the Agronomy Department: that of the position occupied by Dr David M. Van Doren at the Ohio Agricultural Research and Development Center (OARDC) at Wooster, and that of Prof George S. Taylor at the main campus of OSU at Columbus. Rattan Lal was encouraged to consider both positions but he chose the one at the main campus. Whereas the position at Wooster would have been tenured at full professor rather than untenured associate professor, working on the main campus was important to him and his family. His children and wife were able to study at OSU because he was a faculty member. Being on the main campus also attracted visiting scholars from around the world to come to OSU. Indeed, graduate students and scholars from around the world were the main strength of Rattan Lal's academic programme at OSU.

The major challenge of joining OSU was obtaining the U.S. visa. Upon accepting the offer from OSU, Rattan Lal submitted his letter of resignation from IITA in 1984. It was only after he submitted the letter and gave 1 year's notice that he learnt that the wait period for an immigrant visa for applicants from India/South Asia was three or more years. Once again, good fortune prevailed, and the new DG of IITA was extremely helpful, understanding and co-operative. Initially, in March 1987 (more than two-and-a-half years after he had been offered the position), Dr Taylor informed Rattan Lal that OSU, through the Office of International Affairs (Dr Francille Firebaugh), approached the Ohio Senator John Glenn to facilitate his visa. Rattan Lal received a telex from the U.S. Embassy in Lagos at the end of June stating that his visa had been approved and that he must report at a U.S. port of entry by 15 July 1987.

Rattan Lal, his wife and four children (aged 8–13) arrived in New York on 13 July 1987 and reported to the Agronomy Department on 15 July. Thanks to Senator Glenn, 15 July 1987 was the luckiest day of Rattan Lal's professional career. He wanted nothing more than to be a faculty member at his dream university. He never again thought of looking for any other position within or outside OSU. He was finally back where he started from about 20 years before. His arrival at the university and successful adjustment to the LGU system was because of the excellent support that he received both from IITA and the OSU management, staff, faculty and students.

7.4 Support by the Management at IITA and The Ohio State University

Rattan Lal received strong and unconditional support from the management at IITA, both in developing a strong research programme and also in facilitating his transfer from IITA to OSU. Similarly, those who helped him at OSU were Dr George Taylor, Dr Berlie Schmidt, Dr F.P. Miller, Dr G. Mullens, Dr J. Bigham and others at department level; Dr B. Moser, Dr B. McPheron, Dr D. Benfield, Dr S. Slack and Dr D.O. Hansen at college level; and others at university level, as outlined in the acknowledgements section in this book. Rattan Lal received strong support from several presidents of OSU. Dr Kristina

Johnson presided over the launch of the C-FARM project funded by FFAR for a total of $10 million for 5 years. She also provided a match for Rattan Lal's contribution of US$ 1 million to OSU. These funds ($50,000 p.a. for 5 years) were transferred to the professor who occupied the Dr Rattan Lal Endowed Professorship for the first 5 years. Dr and Mrs Michael Drake led the delegation to attend the 2019 Japan Prize in Tokyo. Dr Drake invited Rattan Lal to be a 2019 summer commencement speaker. This is the only time that Rattan Lal was a commencement speaker at the large land grant university. In the provost office, very good support was provided by Dr Bruce McPheron and Dr R. Bellamkonda. It was these well-wishers from OSU and many others in the CGIAR and LGU systems who supported Rattan Lal's work and created an environment for him to be a successful researcher and teacher.

7.5 Teaching at OSU

Rattan Lal's appointment at OSU was 75% research and 25% teaching. This combination was beneficial to him in developing both research and teaching programmes. He was assigned two classes for the first year: Environmental Soil Physics; Soil and Water Conservation. The second class was replaced with the Soil and Climate Class around 2010. Ratan Lal's lectures in both classes were heavy on physics, hydrology, climatology and algebraic calculations. Most students from the SENR and from within the college did not have a strong background in mathematics and basic sciences. Thus, the enrollment was low (10–15 students in environmental physics and 15–20 in soil and climate). In this regard, the department head and director of SENR was always helpful and understanding.

The student evaluation of instructor (SEI) forms, computerized and made online from about 2015, are highly informative and sometimes humorous to read, and anonymous. Examples of some questions and student responses are as follows: 1. What did you like most about the class? A: The day the instructor was absent. 2. What did you like least about the class? A: Math, equations, basic sciences, and too much homework. 3. What changes would you suggest in the curricula? A: Teach agriculture rather than physics and math. 4. Was the instructor interested in the subject matter? A: No, he was interested in using equations and not in teaching agriculture. Despite repeated discussions with the Academic Affairs Committee, this issue of having minimal science and maths requirements remained to be effectively addressed. Rattan Lal had been advised by the Academic Affairs Committee not to use maths in the class. Regardless of eliminating maths and not giving quizzes or exams, Rattan Lal's classes had low enrollment (10–15 in environmental physics and 15–25 in soil and climate) and were under the continuous threat of being cancelled. Naturally, faculty popular with the department and college administration were those which had a high enrollment in their classes. The curricula (time and days of offering the classes) indicated the preferences given by management to classes with high enrollment.

Another option for classes with low enrollment was to offer these in alternate years. The offer of low-enrollment classes in alternate years did not increase the enrollment. Instead, it aggravated the problem because some students graduated without taking these classes, and that was indeed a serious issue because students graduated without adequate knowledge of the specific class. The low enrollment was the basis of Rattan Lal's personal philosophy: a life is better bigger than longer, and a class is better intellectual than bigger.

7.6 The Committee Assignment

There were committees at the department, college and university level to address specific issues such as enrollment, curricula, promotion and tenure. Rattan Lal was fortunate because of a minimal involvement in committee assignment at department, college or university level. After being appointed to the post of Distinguished University Professor, he was eligible (not appointed) to serve on the President and Provost Advisory Committee (PPAC). One important task assigned to him was to serve on the committee that would consider development of Hall of Fame for academic accomplishments at OSU. However, he had some high-level committee assignments at the national and international levels (see Section 8 on International Activities).

7.7 Rewards of Hard Work with Focus on Science

Rattan Lal was successful because he focused on science, teaching and publication. He published his results in books, monographs, book chapters and refereed journal articles. Above all, he travelled widely and presented his ideas at global forum through in-person presentations and webinars via the digital systems. Needless to say, Rattan Lal was known more internationally than in his own department, college or university. It was his global recognition that won him some of the most prestigious awards of the soil and agricultural profession. For example, he received three world prizes (2018 World Agriculture Prize, 2018 World Soil Prize, and 2020 World Food Prize). He also believed that these awards were a recognition of the soil science profession, and that he was merely a representative of the 65,000 soil scientists and received these awards on their behalf.

7.8 International Activities

The OSU had a long history of international programmes at the college and university level. Rattan Lal's arrival and the visa processes were facilitated by these offices. Since his arrival at OSU, he developed international research and teaching programmes at college level by working with Dr David Hansen, and at university level by working with Dr Francille Firebaugh. Dr Hansen helped Rattan Lal towards getting involved in ongoing programmes in Uganda and Tanzania. Rattan Lal, working with Dr Hansen, organized several international conferences and published proceedings of these conferences. They organized visits of three presidents of OSU to India: Dr Kirwan (1999), Dr Holbrook (2000) and Dr Gordon Gee (2009). In 2000, the Ambassador of the USA to India was Dick Celeste, the former governor of Ohio. His younger brother, Ted, was Chair of the Board of Trustees of OSU. Ambassador Celeste invited the Chair of the board and other faculty to India (Fig. 7.1).

With the support and encouragement of the long-term Dean of the College of Food, Agriculture and Environmental Sciences (CFAES), Dr Bobby Moser, Dr Hansen and Rattan Lal also nominated several international scientists and managers for the award of Honorary Degree from OSU. A total of 18 honorary degrees were awarded under the leadership of Dr Moser and Dr Hansen. Among those who received honorary degrees were soil scientists (Kirkham, Gardener, Brady, Greenland, Pretty) and international agriculturists (such as Dr M.S. Swaminathan along with Dr G. Khush, Dr Raj Paroda, Dr Bob Havener, Dr N. Borlaug and Dr Ismail Seraguldin), and philanthropists (Ratan Tata) (Fig. 7.3). The period 1990–2010 was a golden era of the international programme at OSU because of the superb leadership of Dr B. Moser. In addition to Dr Moser and Dr Hansen (and later Dr Luis Canas) at the college level, the Office of International Affairs at OSU provided strong support through Dr F. Firebaugh, Dr J. Ladman, Dr Gil Latz and Dr Kaya Sahin, among others, who also supported these international activities.

7.9 Advances in Soil Sciences

Dr Bobby Stewart, eminent soil scientist with a distinguished career with USDA and later at the Texas A&M University (TAMU), was the founder Editor-In-Chief of the series 'Advances in Soil Science'. He visited IITA in 1986 and was a mentor, colleague and friend of Rattan Lal. From 2010 to 2020, Dr Stewart invited Rattan Lal to work with him as co-Editor-In-Chief. Upon his retirement in 2010, Dr Stewart generously and kindly asked Rattan Lal to continue as Editor-In-Chief of the series. While working with Dr Stewart (Fig. 7.4), Rattan Lal learned more about soil science and about editing and managing a series than ever before.

7.10 Scientific Liaison Officer for USAID to the CGIAR system

Based on Rattan Lal's 18 years of service with IITA and the CGIAR system, Dr Nyle Brady, the Assistant Administrator of USAID in the early 1990s, invited him to his office in Washington and offered him the honorary position of Scientific Liaison Officer (SLO) for Natural Resources Management for 3 years.

Fig. 7.1. Dick Celeste (*front left*), Ambassador of the USA to India in 2000, invited the Chair of the Board of Trustees and Dr Bobby Moser, Vice-President of the Agricultural Administration, to India. He invited the delegation from OSU to his residence in New Delhi and arranged a meeting of the delegation with Ratan Tata (Fig. 7.2). (Photo: Dr David Hansen)

The former SLO was Dr Bobby A. Stewart. To follow in the footsteps of the most internationally recognized soil scientist in the world was both an opportunity and a challenge. Dr Brady explained to Rattan Lal the expectation of the assignment and suggested that he work with the directors general of centres such as CIMMYT, CIAT, ICARDA, ICRISAT and IITA. These are the centres with significant programmes on natural resource management. In addition to visiting the DGs, Rattan Lal would also attend meetings of the Board of Trustees of these centres. Rattan Lal was pleased that the very first visit as SLO was to ICARDA in Aleppo, Syria. This visit was also the last visit as SLO of Dr Stewart to ICARDA. This was a most educational experience for Rattan Lal in terms of the translation of complex science into diplomatic language that was expected by USAID. Just prior to leaving the office of USAID, Dr Brady called Rattan Lal back to advise him about his assignment and stated: 'Rattan, remember that you do not represent USAID, you have no authority, and no budget, other than that the sky is the Limit'. This advice from the greatest agricultural administrator of his time was something that stayed with Rattan Lal wherever he went in his professional assignments.

7.11 Appointments to National and International Committees

During his first year at OSU, Rattan Lal was elected chair of two international organizations: the World Association of Soil and Water Conservation (WASWAC, 1988–1991), and the International Soil and Tillage Research Organization (ISTRO, 1990–1992). As President of ISTRO, Rattan Lal organized the 12th International Conference of ISTRO at IITA, Ibadan, Nigeria, 8–12 July 1991. Following the conference, there was a bus tour from Nigeria to Togo, Benin, Ghana and Niger. The participants flew out of Niamey and returned to their countries.

Fig. 7.2. Dr M.S. Swaminathan (*second from right*) received the honorary degree from the Ohio State University in 1999. Others in the photo are Dr B.Moser and Dr D. Hansen. (Photo: Courtesy of OSU)

7.12 President of the Soil Science Society of America

Rattan Lal was also elected President of the Soil Science Society of America (SSSA) for 2006–2008. During his tenure as President, Rattan Lal invited the President of Iceland, Dr Ólafur Ragnar Grimsson, Fig. 7.5) to be the keynote speaker at the annual meeting of the Tri-Society in 2007. President Grimsson made a virtual presentation, and also invited Rattan Lal to visit Iceland. He and other members of the OSU delegation were invited to a state dinner at the official residence of the President of Iceland. Rattan Lal also invited the President of Bangladesh (who had a PhD in soil science from the University of Wisconsin) to be the keynote speaker at the annual meeting of the Tri-Society in Nashville. During the conference, Rattan Lal introduced President Ahmed to Dr Norman Borlaug (Fig. 7.6). President Ahmed invited Rattan Lal to visit Bangladesh. Along with two other presidents of SSSA, Rattan Lal visited Dhaka, and members of the SSSA delegation were invited to Banga Bhavan for the official state dinner by President Ahmed (Fig. 7.7).

Rattan Lal was privileged and fortunate in being appointed to some key domestic and international appointments. Notable among these were Chair of the Science Advisory Committee of the UNU-FLORES in Dresden, Germany, from 2014 to 2019 (Fig. 7.8), and Chair of the meetings of the Scientific Advisory Committee of UNU-FLORES (Fig. 7.9).

7.13 International Union of Soil Sciences

Rattan Lal was elected President of the International Union of Soil Sciences (2014–2018). It was formerly called the International Society of Soil Sciences (ISSS). His keynote address at the conference in Jeju, South Korea, in 2014 was entitled 'Soil and Peace: Our Common Future' (Fig. 7.10; Lal, 2015). The opening address was attended by about 2500 participants. He also

Fig. 7.3. Ratan Tata (*second from right*) received the honorary degree from the Ohio State University in 1999. Others in the photo are Dr David Hansen (*left*), Director of the international programme in agriculture, and Dr Bobby Moser (*second from left*), Dean of the College of Food, Agriculture and Environmental Sciences at OSU. (Photo: Courtesy of OSU)

presented the opening and closing addresses of the IUSS conference in Rio in 2018.

There were two key activities of IUSS that Rattan Lal presided over in his tenure. One was awarding the Distinguished Service Medal to policy-makers who made a significant contribution to soil science. This programme was started in 2016 and three recipients of the award during Rattan's tenure as President were Stéphane Le Foll of France (Fig. 7.11), Dr Klaus Töpfer of Germany (Fig. 7.12) and Aziz Akhannouch, Minister of Agriculture of Morocco, who was appointed Prime Minister of Morocco (Fig. 7.13). Rattan Lal travelled to Paris, Bonn and Ben Guerir to present these awards.

Rattan took over the position of President of IUSS from Dr Rainer Horn (Germany) and handed it over to Dr Takahi Kosaki (Japan). It was a great pleasure and privilege working with these two world-class professionals (Fig. 7.14). As President of the IUSS, Rattan Lal's participation in IUSS activities were funded by the CFAES. The funding support received from the CFAES was indicative of the strong interest and high priority of CFAES to international activities.

7.14 Assignment with the U.S. Department of Defense as a Member and Chair of the SERDP SAB

Rattan served as a member of the Scientific Advisory Board (SAB) of the Strategic Environmental Research and Development Program (SERDP) of the U.S. Department of Defense from 2011 to 2018. The SERDP SAB is a formal federal advisory committee that has the authority to make recommendations to the SERDP Council regarding technologies, research, and development. Rattan Lal was reappointed to SERDP-SAB from 2023 to 2025 and served as Chair (Fig. 7.15). This was the highest professional honorary appointment that Rattan ever had in the U.S. Government. It was a great pleasure working with Jon Bunger and the staff of SERDP-SAB who made arrangements

Fig. 7.4. Dr Bobby Stewart visited IITA in 1986 and Rattan Lal had the honour to host his visit and later worked with him as co-Editor-in-Chief of the series 'Advances in Soil Science' in 2020. Dr Stewart was a professional mentor and key advisor of Rattan Lal. (Photo: Lal Carbon Center)

for travel and handled logistics. Members of the SAB were highly competent professionals in their areas and Rattan learned many new things about scientific disciplines other than soil science and agronomy.

7.15 Board for International Food and Agriculture Development

Rattan Lal was appointed a member of the Board of International Food and Agriculture Development (BIFAD) by President Joe Biden from 2022 to 2025 (Fig. 7.16). This was a high-level professional appointment and it was a great pleasure to work with Dr Rob Bertam, Clara Cohen and other staff of USAID.

Being an appointee of President Biden, Rattan Lal was invited to a farewell reception for him at the White House in January 2025. However, because of teaching commitments, Rattan Lal did not attend the reception. However, he had the great honour and privilege to meet and listen to the presentations of the USAID administrator and other staff.

The historic event of the closing of the U.S. Agency for International Development in January 2025 automatically led to the cessation of BIFAD and its activities.

7.16 International Fertilizer Development Center (IFDC)

The IFDC, based in Muscle Schoals, Alabama, is an independent non-profit organization that combines innovative research, market systems development, and strategic partnerships to spread sustainable agricultural solutions for improved soil health, food security and livelihoods around the world. Its mission is healthier soils and plants for food and environmental security. Its research and development programme is aimed at promoting innovative research on managing

Fig. 7.5. In 2006, Rattan Lal was invited to the University of Iceland in Reykjavik and had an audience at the official residence of the Icelandic President Dr Ólafur Ragnar Grímsson. President Grimsson also was a keynote speaker for the Tri-Society when Rattan was President of SSSA in 2007. However, President Grimsson presented his lecture online. (Photo: Iceland President's Office)

plant nutrients that benefits farmers and land managers. Rattan Lal served as a member of its Board of Directors from 2020 to 2026, and his professional goal was to enhance the fertilizer-use efficiency by improving soil health, decreasing losses and observing the law of return.

7.17 Inter-American Institute of Cooperation in Agriculture (IICA)

IICA, based in San José, Costa Rica, is aimed at improving agriculture of the Inter-American system in co-operation with its member states.

In 2020, Rattan Lal was appointed Chair of Soil Science and Goodwill Ambassador for Sustainable Development Issues of IICA. He found IICA to be a great international centre to work and one which had a broad-based mandate to make improved (science-based) agriculture a part of the solution for restoring soil health and improving the environment.

He worked closely with its management including Dr Manuel Otero, Dr Lloyd Day, Dr Jorge Werthein and Dr Muhammad Ibrahim (Fig. 7.17). Rattan Lal participated in COP27, 28, 29 and 30 as a member of the IICA team. His guide and advisor at these and other meetings was always Dr Francisco Mello, and Rattan Lal greatly appreciated his co-operation and support. He served in this position for 6 years and addressed ministerial conferences involving secretaries of agriculture and environment from 34 countries of the Americas.

Fig. 7.6. The 2007 ASA CSSA Soil Science Society of America President introducing the President of Bangladesh, His Excellency Dr Iajuddin Ahmed, to Dr Norman Borlaug at the annual Tri-Society meeting of SSSA hosted in Nashville, Tennessee. Dr Iajuddin Ahmed holds a PhD in soil science from the University of Wisconsin. Rattan Lal also visited the Banga Bhavan in 2009. (Photo: Tri-Society: ASA/CSSA/SSSA, Madison, Wisconsin)

Fig. 7.7. In 2009, the President of Bangladesh, Dr Iajuddin Ahmed, invited Rattan Lal to a state dinner at Banga Bhavan. (Photo: Rattan Lal personal photo album)

Fig. 7.8. Rattan Lal with Dr David Malone, Rector of the United Nations University, Tokyo, and Dr Karl-Heinz Feger, Dean of the College of Agriculture and Forestry of the Technical University of Dresden. Dr Malone appointed Rattan Lal as Chair of the Scientific Advisory Committee of UNU-FLORES in Dresden. (Photo: UNU FLORES, Dresden, Germany)

Fig. 7.9. Rattan Lal chaired the meetings of the Scientific Advisory Committee of the UNU-FLORES at its headquarters in Dresden, Germany. Prof Karl-Heinz Feger, Dean of the College of Agriculture and Forestry at TUD, is on the far left of the picture. (Photo: UNU-FLORES, Dresden, Germany)

Fig. 7.10. Keynote presentation by Rattan Lal at the opening session of the International Union of Soil Sciences in 2014 at Jeju, South Korea, when he was elected President of the IUSS. (Photo: IUSS)

Fig. 7.11. The distinguished Service Medal Award of IUSS was presented by Rattan Lal to His Excellency Stéphane Le Foll, Minister of Agriculture and Forestry in France. The Minister also visited Ohio State at the end of June 2015 to discuss the launch of the 4 per 1000 programme of COP21 in Paris. Later in the day, Le Foll also visited the no-till farm of David Brandt. Le Foll was impressed by seeing cover cropping and no-till farming practised by Brandt and talked to farmers about practical applications of conservation agriculture. (Photo: IUSS)

Fig. 7.12. In 2019, the Distinguished Service Award was presented by Rattan Lal to Dr Klaus Töpfer, CEO of the Institute for Advanced Sustainability Studies in Potsdam, Germany, during Global Soil Week in Bonn, Germany. He was also Secretary of the Environment of Germany when Helmut Kohl was the Chancellor. Others pictured include: Dr Joachim Von Braun (*far right*); Margret Thalwitz (*second from right*), Chair of Executive and Finance Committee of ICARDA; Dr Rainer Horn (*centre, behind award*), President of IUSS before Rattan Lal; and Dr Takashi Kosaki (*second from left*), President of IUSS after Rattan Lal. (Photo: IUSS)

Fig. 7.13. As President of the International Union of Soil Science from 2018 to 2019, Rattan Lal presented the Distinguished Service Medal to His Excellency Mr Aziz Akhannouch, the Prime Minster of Morocco, in 2019. This award was presented to recognize his role in implementing a programme called Adaptation of African Agriculture (AAA) Initiative launched in 2016 at the UNFCC meeting in Marrakesh, Morocco. (Photo: IUSS)

Fig. 7.14. Three presidents of IUSS: Dr Rainer Horn (*right*; Past President); Rattan Lal (*centre*; President); and Takashi Kosaki (*left*; incoming President). (Photo: Takashi Kosaki)

Fig. 7.15. Oath ceremony for appointment as Chair of the SERDP SAB of the U.S. Department of Defense in 2023. (Photo: SERDP SAB of DOD, Washington)

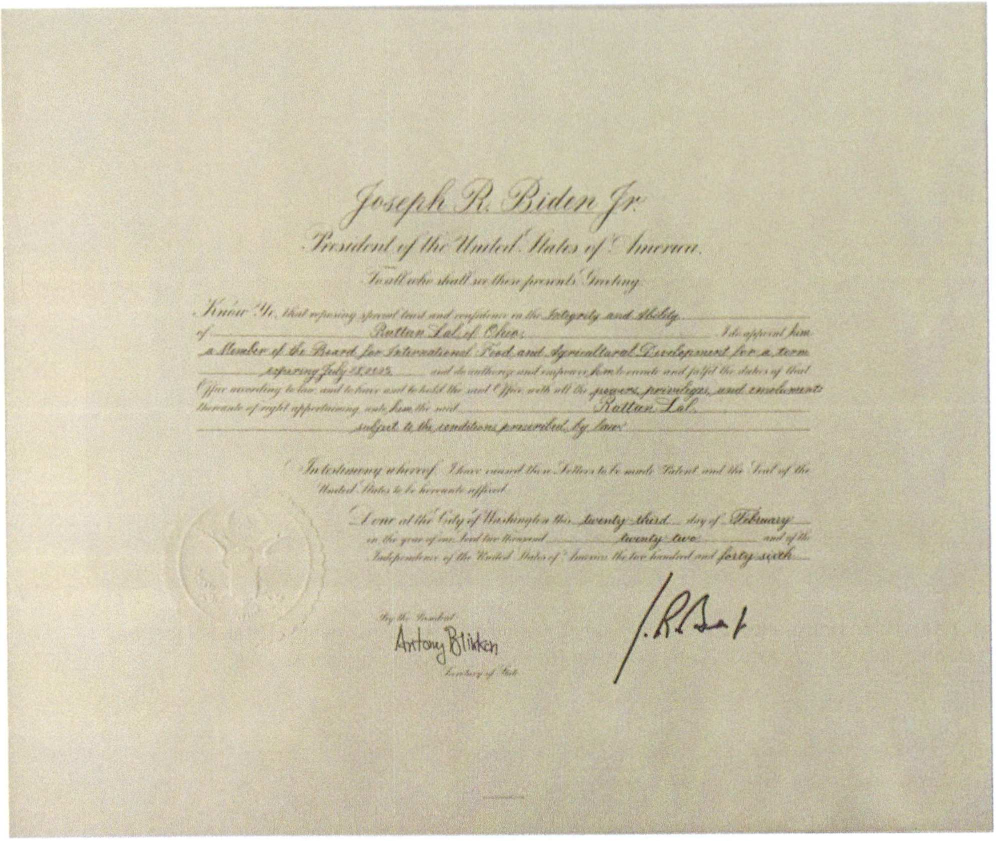
Joseph R. Biden Jr.
President of the United States of America
To all who shall see these presents, Greeting:
Know Ye, that reposing special trust and confidence in the Integrity and Ability of Rattan Lal of Ohio, I do appoint him a Member of the Board for International Food and Agricultural Development for a term expiring July 28, 2025, and do authorize and empower him to execute and fulfil the duties of that Office according to law, and to have and to hold the said Office with all the powers, privileges, and emoluments thereunto of right appertaining, unto him, the said Rattan Lal subject to the conditions prescribed by law.
In testimony whereof, I have caused these Letters to be made Patent and the Seal of the United States to be hereunto affixed.
Done at the City of Washington this twenty third day of February in the year of our Lord two thousand twenty two and of the Independence of the United States of America the two hundred and forty sixth
By the President
Antony Blinken
Secretary of State

Fig. 7.16. Appointment of the BIFAD members was made by President Joe Biden (see Chapter 22, Section 10). Photo: Lal Carbon Center.

Fig. 7.17. The high-level management team of IICA in 2022 with whom Rattan worked in implementing its programme and advancing its mission throughout the Americas and globally. Those pictured (from right to left) are: Manuel Otero, Rattan Lal, Jorge Werthein and Lloyd Day. Dr Muhammed Ibrahim (*right*) joined IICA in 2024 as Director for Technical Cooperation. (Photo: IICA)

Bibliography

Da Vinci, L. (n.d.) *Codex Atlanticus*, folio 680 (recto). Biblioteca Ambrosiana, Milan, Italy.

Gabor, D. (1963) *Inventing the Future*. Secker and Warburg, London.

Gates, B. (2023) CGIAR at COP28: Ground breaking innovative work.

Lal, R. (1987) *Tropical Ecology and Physical Edaphology*. John Wiley & Sons, Chichester, UK.

Lal, R. (2015) The soil-peace

nexus: Our common future. *Soil Science and Plant Nutrition* 61, 566–578.

Lele, U. (2021) Moving towards "One CGIAR". *International Journal for Rural Development*, 11–13.

Nye, P.H. and Greenland, D.J. (1965) *The Soil Under Shifting Cultivation*. Farnham Royal, Bucks., Commonwealth Agricultural Bureaux.

8 Vegetarian Diet and Home Gardening

'So I am living without fats, without meat, without fish, but am feeling quite well this way. It always seems to be that man was not born to be a carnivore.'
— Albert Einstein, in a letter to Hans Muehsam in 1954

'Show me your garden and I shall tell you what you are like.'
— Alfred Austin

'To plant a garden is to believe in tomorrow.'
— Audrey Hepburn

Abstract
Rattan Lal has followed a vegetarian diet, along with being teetotal and a non-smoker, for all his life. In addition to the influence of his bebe, sister and bua ji, who were all strictly vegetarian, Rattan Lal was also influenced by his fourth-grade teacher who was a strong mentor and follower of Vedic traditions. In addition to health benefits, a vegetarian diet and simple lifestyle have a low environmental or ecological footprint. The projected increase in the world's population by another 1.5 billion by 2050 could have a strong impact on demand for land and water. Thus, widespread adoption of a vegetarian diet can spare land and water for nature and make agriculture and food systems a part of the solution to address global issues of the modern era. The goal of agriculture is to produce more from less land, water, use of chemicals and energy, and spare about half of the land under agriculture for nature conservancy. Home gardening is also an important tool to transform the world food systems. The slogan should be 'One home, one vegetable garden'.

8.1 Food Habits

Rattan Lal was brought up in a vegetarian family following the Vedic culture and tradition. He maintained vegetarian dietary habits throughout his life. In addition to being a vegetarian, he is also a teetotaler and a non-smoker, and has never used any substances. He does not drink coffee. When he became familiar with global issues, he also understood that the vegetarian food habits may also be healthy for both the planet and the people. But, diet being a personal choice, it has rather become a debatable issue. It is therefore important to understand different types of diet and their environmental (land, water, climate) footprint so that one can make an objective and rational decision about dietary habits that are good for human and environmental issues. It is also important to realize that credible data on environmental footprint are scarce. From whatever is available in the popular literature, it is difficult to decide what is correct and what is not. Thus, some examples of the environmental footprint of different types of food are just examples and should be used with caution. Scaling up from these examples to global scale for the entire human population of the world may not be accurate.

8.2 Types of Diet

There are several types of diet based either on religion or source of the food. Based on the religion, diet is classified on beliefs as follows:

DOI: 10.1079/9781836990154.0008

1. The Hindu diet restricts intake of animal-based food except dairy products (e.g. milk, cheese, butter, yoghurt). It involves the use of whole grains, vegetables, fruits, nuts and dairy products. It is a diet that comes from all sources but does not involve killing of animals. This transformation of Hindus to a vegetarian diet happened during the Buddhist era about 600 BCE.
2. The Buddhist diet is primarily vegetarian for the Mahayana sect.
3. The Jain diet is strictly vegetarian, because of their belief in *ahinsa* (no harm to any animal). However, followers of Jainism may be lacto-vegetarian or completely vegan. Some Jains also restrict consumption of root vegetables because harvesting of roots may harm soil animals (e.g. earthworms, termites, microbes etc.).
4. The Islamic diet does not restrict consumption of animal-based food but animals may be slaughtered in different ways. Islamically impermissible food may not include alcoholic beverages, pork etc.
5. The Kosher diet is based on laws of Judaism, which discourage meat consumption and prohibit alcoholic or caffeinated beverages.

Human diet may also be classified based on the source of food such as plants or animals. Based on the source of the food, there are several types of human diets as follows:

1. The vegan diet does not include any animal products (e.g. dairy products, eggs, fish, poultry, meat). It primarily involves whole grains, fruits, vegetables and nuts.
2. The vegetarian diet is a plant-based diet involving whole grains, vegetables, fruits, nuts, dairy products (milk, cheese, yogurt, butter milk). Vegetarians may be ovo-vegetarian – consume eggs – or ovo-lacto vegetarian – consume both eggs and dairy products.
3. The fruitarian diet consists of using raw fruits only. Fruitarian diet is also used during some religious fasts.

The above three diets are primarily classed as plant-based diets (PBDs).

4. The flexitarian diet is a semi-vegetarian diet that emphasizes plant-based foods while occasionally including meat or other animal products.
5. The western diet refers to highly processed food, including red meat (beef, pork, bacon), saturated and trans fats, butter, cheese and fried foods, and high-energy sugary drinks.

Rattan Lal has practised the vegetarian diet (whole grains, pulses, vegetables, nuts, fruits, dairy products) from childhood.

8.3 Dietary Effects on Human Health

Diet can have strong effects on human health, but it is difficult to generalize these effects because of the interaction of diet with environment and genetic or heredity factors. All other factors (environment and genetic) remaining the same, PBDs may be healthier than animal-based diets. In general, PBDs (whole grains, nuts, vegetables and fruits) may reduce risks of disease such as cancer, coronary heart disease, birth defects, cataracts, and obesity (Willett, 1994).

Therefore, there is a potential for disease prevention in the population using western diets by changing to PBDs. Similarly, intake of dietary fibres can have protective effects against some gastrointestinal diseases such as constipation, haemorrhoids, colon cancer, ulcers, diabetes, stroke, hypertension and cardiovascular disease (Otles and Ozgoz, 2014). There are also beneficial effects of dietary phospholipids on human health (Gaudry *et al.*, 2012). Whereas specific impacts of diet on human health are difficult to identify because of the interaction of environment and genetics, PBDs can be beneficial in preventing and curing several human diseases.

8.4 Dietary Effects on Environment

There are several global issues which are related to effects of the growing and increasingly affluent human population on the environment (see Preface). Important among these are the growing risks of anthropogenic climate change, being aggravated by emission of greenhouse

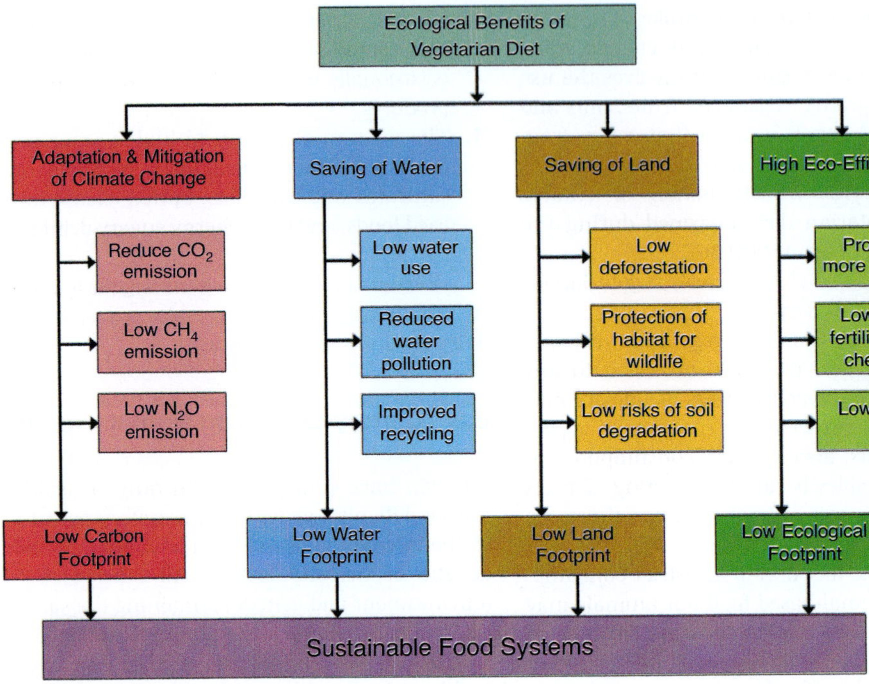

Fig. 8.1. Some ecological benefits of vegetarian, vegan and other plant-based diets. Saving in land and water and returning some of it to nature is an important consideration. (Adapted from Lal, 2024)

gases (GHGs), water scarcity and pollution, loss of biodiversity, soil degradation etc. Similar to the effects of diet on human health because of the interaction with genetics and the environment, it may be difficult to establish the direct cause–effect relationship between diet and the risks of environmental degradation. Yet it is generally argued that PDBs have a lower environmental footprint than that of predominantly animal-based or western diets.

Some benefits of the PBD to the environment in Fig. 8.1 show potential reduction in carbon footprint, water footprint, land footprint and ecological footprint. The magnitude of these effects may vary widely due to difference in site-specific situation.

8.5 Sparing Land by Adoption of Plant-Based Diet

It is important to understand that modern intensive farming has been successful in producing enough food from the same land to feed a growing population of 8.2 billion in 2024 (UK government, 2024). Yet, humans are already using 50% of the Earth's ice-free and desert-free land area for agriculture and food production. It is also widely believed that there is a strong need to reduce the land footprint of humanity. The rate of anthropogenic CO_2 increase in the atmosphere is >2 ppm per annum (Friedlingstein, 2024), and it needs to be reduced if global warming is to be kept below 2°C by the end of the century. Thus, an urgent issue of global significance is to harness the potential of shifting to PBD with the goal of reducing land and water footprint of humanity and sparing some land for nature.

As outlined in Table 8.1, there is a lot of interest in sparing land for nature by adoption of PBDs. It is argued that raising livestock (for meat and dairy) requires more land and water than that for growing PBDs. The land area needed to produce 1000 kilocalories of feed ranges from < 4 m² for PBDs (maize, rice, wheat,

Table 8.1. Land use of foods per 1000 kilocalories. (Recalculated from World in Data, 2024, based on Poore and Nemecek, 2018)

Food	m^2 of land	Relative
Beef (beef herd)	119.5	170.7
Lamb and mutton	116.7	166.71
Cheese	22.7	32.43
Beef (dairy herd)	15.8	22.57
Milk	14.9	21.29
Pig meat	7.3	10.42
Poultry meat	6.6	9.43
Fish (farmland)	4.7	6.71
eggs	464	6.29
Tomatoes	4.2	6.0
Bananas	3.2	4.57
Oatmeal	2.9	4.14
Prawns (farmed)	2.9	4.14
Citrus fruit	2.7	3.86
Peas	2.2	3.14
Nuts	2.1	3.0
Cassava	1.9	2.71
Groundnut	1.6	2.29
Wheat and rye	1.4	2.0
Apples	1.3	1.86
Potatoes	1.2	1.71
Root vegetables	0.9	1.29
Rice	0.8	1.14
Maize	0.7	1.0

Table 8.2. How much water is needed to produce 1 kg of food. (Adapted from Institution of Mechanical Engineers, n.d.)

Food	Litres of water/kg	Relative
Tomato	214	1
Cabbage	237	1.11
Brussels sprout	283	1.32
Potato	287	1.34
Orange	533	2.49
Banana	790	3.69
Apple	822	3.84
Peach	910	4.25
Rice	2497	11.67
Tofu	2513	11.74
Cheese	3178	14.83
egg	3920	18.32
Chicken	4325	20.21
Butter	5553	25.95
Pork	5998	28.03
Sheep	10,412	48.65
Beef	15,000	70.01

vegetables) to 4–7 m^2, compared to 7–20 m^2 for dairy products and 115–120 m^2 for red meat from lamb/mutton and beef. With baseline of land required for 1000 kilocalories of maize as 1, the relative land required for red meat is about 120 (Table 8.1). It has thus been estimated that if everyone were to switch to PBDs, it would spare as much as 75% of the land currently used for agriculture (5.2 billion ha). Vegan diet can be produced on 1 billion ha of land compared with 4.1 billion ha required to feed the current population with the diet habits of the 2020s. This is an academic exercise because it is difficult to change the food habits of the population. Indeed, there is a growing demand for animal-based diets as the people of developing countries (China, India, Mexico, Brazil, Russia etc.) become rich and more affluent.

8.6 Sparing Water by Adopting PBDs

Similar to savings in land, adoption of PBDs can also save water. The water footprint of people living in different continents varies widely from 2617 l/d capita in North America to 3227 l/d capita in Europe. However, total population varies widely among continents with the highest being in Asia. The water footprint also varies among different diet indicators. The data in Table 8.2 show differences in the water footprint according to different food indicators. Total water footprint is 6056 l/person/d which includes green water (absorbed from soil to grow plants) at 4950 l/person/d, blue water (water from rivers, lakes, groundwater, sea etc.) at 535 l/person/d and grey water (water contaminated by human

Table 8.3. Land area required to feed one person in the USA for 10 diet scenarios. Total persons fed/ha are calculated from the data in the last column. (Adapted and recalculated from Peters *et al.*, 2016)

	Hectare				
Diet scenario	Cultivated cropland	Perennial cropland	Grazing land	Total land	Persons/ ha
Baseline (BAS)	0.18	0.16	0.74	1.08	0.93
Positive control (POS)	0.15	0.17	0.71	1.03	0.97
100% healthy omnivorous (OMNI100)	0.15	0.14	0.64	0.93	1.08
80% healthy omnivorous (OMNI80)	0.14	0.10	0.55	0.79	1.27
60% healthy omnivorous (OMNI60)	0.13	0.07	0.45	0.65	1.54
40% healthy omnivorous (OMNI40)	0.13	0.05	0.28	0.46	2.17
20% healthy omnivorous (OMNI20)	0.12	0.03	0.10	0.25	4.0
Ovolacto vegetarian (OVO)	0.12	0.02	0	0.14	7.14
Lactovegetarian (LAC)	0.12	0.02	0	0.14	7.14
Vegan (VEG)	0.13	0	0	0.13	7.69

*Persons fed/ha in the last column are calculated by dividing 1 ha by the land area needed to feed 1 person for different diet scenarios.

use and recycled) at 463 l/person/d. Therefore, human population is projected to increase from 8.2 billion in 2024 to 10.3 billion by the 2080s and decline to 10.2 billion by 2100 (UN, 2024). Thus, increase in world population by another 2 billion by 2080 could have a strong impact on world water demand

Similar to land use, there is also a strong impact of diet on the water footprint. A vegan or PBD generally requires significantly less water compared to a diet heavy in meat. Taking the baseline of water use for a vegan diet at 1, the water footprint is estimated at 1.73 for a low-meat diet and 2.17 for the high-meat diet.

The differences in water footprint among diets is attributed to the differences in litres of water needed to produce 1 kg of food from different sources. The water footprint (litre of water per kg of food) is 214 to 287 for vegetables, 533 to 910 for fruits, 2500 for rice and tofu, 3173 for cheese, 3923 to 4325 for eggs and chicken, 5553 for butter, 6000 for pork, 10,412 for sheep and 15,000 for beef (Table 8.2). The relative water footprint (relative to tomato) ranges from 1 to 70. If these data are nearly correct, a lot of saving can be had in shifting diets in favour of PBDs.

8.7 How Many People Can Be Fed from Different Types of Diet Grown on One Hectare of Land

Based on statistics shown in this chapter, the number of people that can be fed from 1 ha of land varies greatly. Ritchie (2021) estimated that we would reduce global agricultural land use from 4 billion ha to 1 billion ha if the world adopted the plant-based diet (Table 8.3). Peters *et al.* (2016) estimated the carrying capacity of the U.S. agricultural land for 10 diet scenarios, and observed that annual per capita land requirement ranged from 0.13 to 1.08 ha/person/year across the 10 diet scenarios (Table 8.3). Peters *et al.* (2016) estimated that carrying capacity of the U.S. agricultural land ranged from 402 to 807 million persons, which was 1.3 to 2.6 times

Table 8.4. A hypothetical case of reducing land area and water use by improving use efficiency of inputs. Producing more from less is critical to returning some land to nature. (Adapted from Lal, 2022, 2023)

Resource Use	Units	2020	2030	2050	2100
Fertilizer	Mt	200	150	100	50
Irrigated Area	Mha	350	400	450	500
Cropland Area	Mha	1500	1400	1000	750
Conservation Agriculture	Mha	200	250	500	750
Grazing/Rangeland	Mha	3700	3500	2500	1500
Water Use for Agriculture	km^3	3150	3000	2000	1000
Global Cereal Yield	Mg/ha	4.00	4.5	6.00	8.00

the U.S. population in 2010. The last column in Table 8.3 is calculated from the data by Peters *et al.* (2016) by dividing 1 ha by the land area needed (person/ha) for different food scenarios listed in column 1. Therefore, the number of people that can be fed for each hectare of agricultural land ranges from 7.7 for the vegan diet to 7.1 for the vegetarian diet, to 0.93 for baseline western diet of the USA. While the upscaling and extrapolation of the data shown in Table 8.3 must be done with caution, there is some commonalities of these calculations with those of Ritchie (2021) shown in Table 8.2. Assuming that vegan and vegetarian diets can feed, on average, of 7.3 persons/ha, feeding a global population of 10.2 billion by 2100 can be done by 1.4 billion ha of agricultural land.

The data in Table 8.4 are based on the assumption that sustainable management of soil and agriculture, reducing food waste and changing to PBD globally can save land and water for nature. If, by 2100, the cropland area can be reduced to 750 Mha and the grazing land area to 1500 Mha (with total land area under agriculture to 2250 Mha), this would save a lot of land area which can be returned to nature for sequestration of atmospheric CO_2 and strengthening of biodiversity. Conversion of flood-/sprinkler-based irrigation to drip irrigation can save a lot of water from 3150 km^3 being used presently to 1000 km^3 by 2100. Restoration of soil health and increasing use efficiency of fertilizers can reduce the total amount of fertilizer use from 200 million tons in 2024 to 50 million tons by 2100. Similarly, improvement of soil health and the attendant improvement in disease-suppressive attributes can lead to reduction in pesticide use. However, it is also assumed that most of the cropland area (750 Mha) will be managed by system-based conservation agriculture (CA) practices (Fig. 8.1).

Adoption of CA can reduce risks of soil erosion and water run-off and emissions of GHGs from agro-ecosystems. Thus, adoption of science-based farming (especially in developing countries of Africa, South Asia, Central America and the Caribbean) will increase cereal grain yield from 4 Mg/ha to 8 Mg/ha and reduce the land area needed for food production.

Urban farming, based on soil-less culture and recycling of grey and black water, and sky farming (multi-story greenhouses to grow vegetables), can be an important strategy to feed megacities while sparing land and water for nature.

8.8 Health Benefits of Plant-based Diets

In addition to reducing the environmental footprint (sparing land and water), adoption of PBD is also claimed to have some benefits to human health (Fig. 8.2). Some popular reports indicate that adoption of a vegetarian diet can have several health benefits such as decreasing risks of non-infectious chronic diseases such as obesity, coronary heart disease, hypertension or high blood pressure, diabetes and some cancers. Eating a plant-based diet can improve the health of gut so vegans and vegetarians are better able to absorb the nutrients from food that support the

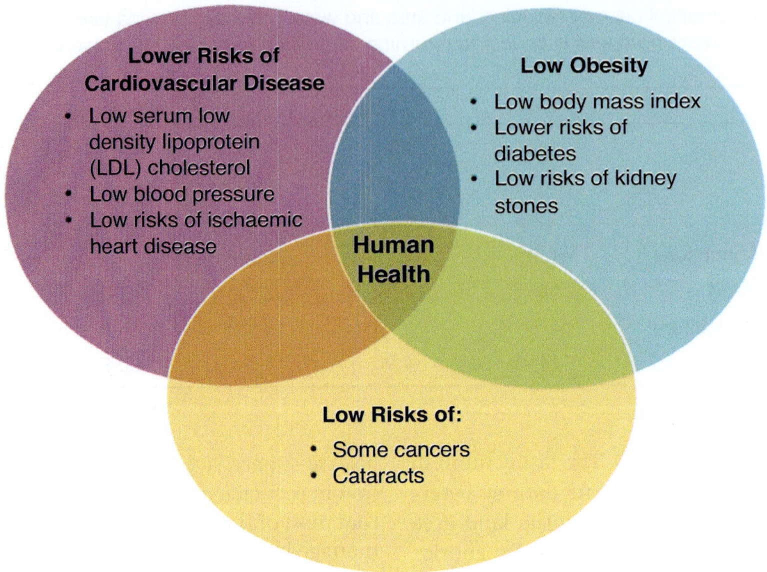

Fig. 8.2. A vegetarian diet can be beneficial to human health by reducing risks of cardiovascular diseases, obesity and diabetes along with those of some cancers and cataracts. (Figure drawn from the information contained in articles by Key *et al.*, 2022 and Wang *et al.*, 2023)

immune system and reduce inflammation. Fibre can lower cholesterol and stabilize blood sugar and it is great for a good bowel. However, not all plant-based foods are equally healthy (Wang *et al.*, 2023). Unhealthy vegetarian diets poor in specific nutrients (vitamin B12, iron, zinc and calcium) and/or rich in highly processed and refined foods increase morbidity and mortality. Key *et al.* (2022) reported that important characteristics of plant-based diets, which would be expected to be beneficial for long-term health, are low intake of saturated fat and high intake of dietary fibre. Key and colleagues reported that of vegan and vegetarians, vegans typically have lower body mass index, serum low-density lipoprotein (LDL) cholesterol and blood pressure than comparable regular meat-eaters, as well as lower bone mineral density. In the EPIC Oxford study, Key *et al.* reported that vegetarians have a relatively low risk of ischaemic heart disease, diabetes, diverticular disease, kidney stones, cataracts and possibly some cancers. On the other hand, potentially deleterious characteristics of vegetarian diets are the risk of low intakes of some micronutrients such as vitamin B12, vitamin D, calcium and iodine, particularly in vegans. Key and colleagues observed that vegetarians and vegans have a relatively high risk of stroke (principally haemorrhagic stroke) and bone fractures, in comparison with meat-eaters. It is widely believed that some nutrients, such as protein, iron, calcium, zinc, vitamin B12 and vitamin D can be harder to obtain from some types of vegetarian diets. Therefore, under the guidance of a nutritionist, a well planned diet (combination of pulses, grains, fruits and vegetables, nuts) can meet nutritional needs during all stages of life including childhood, nursing mothers, and elderly citizens. Nonetheless, special care needs to be taken during pregnancy, breastfeeding and with children for those on vegetarian and, especially, vegan diets. These risks may be mitigated by optimal food choices, fortification and supplementation (Key *et al.*, 2022; Wang *et al.*, 2023).

8.9 Vegetarian Diet and Health of Rattan Lal

It is important to recognize that genetics also play an important factor in human health. As an example, some comments are pertinent about Rattan Lal's health in general. Rattan Lal, being a vegetarian all his life, had overall good health throughout his life. However, he had some inherited cardiac issues which run in the family. His father, brother and one nephew died of heart-related issues. Rattan Lal was fortunate to have access to medical facilities that his father, brother and nephews did not. In spite of all these, Rattan Lal had been advised by his physician to take a daily supplement of multivitamins and an additional dose of zinc and iron. He has followed strictly the advice of his doctors including primary healthcare and cardiologists. Despite being vegetarian, Rattan has been on medicines for treatment of cardiovascular disease since his mid-30s (see Chapter 14).

8.10 Home Garden and Its Impact on Human Health and Well-being

In the context of diet, it is pertinent to state that having a home garden has some positive effect on human health and well-being. Growing up on a farm, and harvesting okra and cantaloupe from a few plants scattered within the cotton plots, was an experience that became a hobby of Rattan Lal's since early 1970 when the family moved from an apartment to a two-bedroom, and then to a 3-bedroom, house at the campus of the International Institute of Tropical Agriculture (IITA). At IITA, Rattan Lal was interested in growing some Asian vegetables such as bottle gourd, which would climb on the roof, and garden shrubs and trees. When the family moved to Columbus, Ohio, in 1987, Rattan took a keen interest in developing a small garden plot (20x 10 ft to 30x 30 ft) in their custom-made home since 2017. Both homes were in Upper Arlington, and a neighbour in the second home complained to the Upper Arlington municipality about some suspicious plants being grown 'on a pile of dirt with rotting wheat straw scattered on the top'. The authorities inspected the plots without informing Rattan Lal, wrote a letter about the complaint and demanded a response within 1 week. When Rattan Lal called the officer who had visited the garden plot, he asked him about his findings. The officer replied that growing vegetables and flowers in the home garden is permitted, but the views of all residents must be respected.

Rattan Lal usually started working in the garden in early spring (mid-April) and mostly transplanted seedlings by 15 May. Some of the seedlings were purchased from the nursery shops (e.g. tomato, bell pepper, chili pepper, cucumber, zucchini, two types of egg plant (round and long)) and others were grown from the seedlings planted at home (e.g. ridge gourd, bottle gourd, bitter gourd, squash) and some were seeded direct. Of the latter, okra was seeded in April, radish and turnips in August, and garlic cloves planted in early November. In 2024, when Rattan had a busy travel schedule from mid-November to mid-December, he planted garlic on 26 and 27 December.

In general, Rattan Lal grew a dozen or more vegetables from mid-spring to late fall. These were tomato, cucumber, zucchini, bell pepper, chilli pepper, e.gg plant, okra, French bean, string bean, potato, garlic, daikon radish, turnip, garlic, onion. Figures 8.3 and 8.4 show Rattan Lal with some of his proud weekly harvest during mid-summer. In some years, he was also successful in growing bitter gourd (Fig. 8.5).

Hobby aside, having a home garden has many practical implications (Lal, 2020). During the COVID pandemic (2019–2020), home garden was an important source of fresh vegetables (Fig. 8.3). Some voluntary garden plants were a great surprise and a useful source of vegetables and fruits. These were apparently spread by seeds in birds' droppings. In addition to cherry tomato, which grew as weeds and often took over the garden and had to be pulled out, voluntary plants were a surprise gift of Mother Nature.

A garden is good for the mind, body, family and health of the urban environment. The slogan of 'One home, one family garden' should be universally promoted.

Fig. 8.3. Rattan Lal's hobby is growing vegetables in his home garden. He grows as many as 12 different vegetables and most do very well. It requires some irrigation during the dry spells. Most vegetables are tolerant to native insects and pathogens. The major problem is rabbit and deer. Thus, a fence is critically important. (Photo: Sukhvarsha Lal)

Fig. 8.4. Lauki or the bottle gourd (*Lagenaria siceraria*) is a favourite vegetable of Asia, and it grows very well in Ohio. (Photo: Sukhvarsha Lal)

Fig. 8.5. Bitter gourd (*Memordica charantia*) has the name of bitter melon, karela, balsam pear, bitter apple and bitter squash. It is an important vegetable for Asia, Africa and the Caribbean and it has also some medicinal uses. (Photo Sukhvarsha Lal)

Bibliography

Einstein, A. (1954) Letter to Hans Muehsam.

Friedlingstein, P., O'Sullivan, M., Jones, M.W., Andrew, R.M., Hauck, J. *et al.* (2024) Global carbon budget 2024. *Earth System Science Data Discussions*, 2024, 1–133.

Gaudry, D., Taylor, L., Schneider, M. and Massing, U. (2012) Health effects of dietary phospholipids. *Lipids in Health and Disease* 11(3). DOI: 10.1186/1476-511X-11-3.

Harris, F., Moss, C., Joy, E.J.M., Quinn, R., Scheelbeek, P.F.D. *et al.* (2020) The water footprint of diets: A global systematic review and meta-analysis. *Advances in Nutrition* 11(2), 375–386. DOI: 10.1093/advances/nmz091.

Institution of Mechanical Engineers (n.d.) How much water is needed to produce food and much do we waste? *The Guardian*. Available at: https://www.theguardian.com/news/datablog/2013/jan/10/how-much-water-food-production-waste (accessed 2 June 2024).

Key, T., Papier, K. and Tong, T. (2022) Plant-based diets and long-term health: Findings from the EPIC-Oxford study. *Proceedings of the Nutrition Society* 81, 190–198. DOI: 10.1017/S0029665121003748.

Küllenberg de Gaudry, D., Taylor, L., Schneider, M. and Massing, U. (2012) Health effects of dietary phospholipids. *Lipids in Health and Disease* 11(3). DOI: 10.1186/1476-511X-11-3.

Lal, R. (2020) Home gardening and urban agriculture for advancing food and nutritional security in response to the COVID-19 pandemic. *Food Security* 12(4), 871–876. DOI: 10.1007/s12571-020-01058-3.

Lal, R. (2022) Returning land to nature. *International Journal for Rural Development*. https://www.rural21.com/fileadmin/downloads/2022/en-02/rural2022_02-S04-07.pdf

Lal, R. (2023) Farming systems to return land for nature: It's all about soil health and re-carbonization of the terrestrial biosphere. *Farming System* 1(1), 100002. DOI: 10.1016/j.farsys.2023.100002.

Lal, R. (2024) Soil degradation effects on human malnutrition and undernutrition. *Medical Research Archives* 12(11). DOI: 10.18103/mra.v12110.5840.

Lares-Michel, M., Housni, F.E., Cervantes, V.G.A., Carrillo, P., Nava, R.M.M. *et al.* (2021) Eat well to fight obesity...and save water: The water footprint of different diets and caloric intake and its relationship with adiposity. *Frontiers in Nutrition 8*. Available at: https://www.frontiersin.org/journals/nutrition/articles/10.3389/fnut.2021.694775 (accessed 2 June 2025).

Otles, S. and Ozgoz, S. (2014) Health effects of dietary fiber. *Acta Scientiarum Polonorum. Technologia Alimentaria* 13, 191–202. DOI: 10.17306/J.AFS.2014.2.8.

Peters, C., Picardy, J., Darrouzet-Nardi, A., Wilkins, J., Griffin, T. *et al.* (2016) Carrying capacity of U.S. agricultural land: Ten diet scenarios. *Elementa: Science of the Anthropocene* 4, 000116. DOI: 10.12952/journal.elementa.000116.

Poore, J. and Nemecek, T. (2018) Reducing food's environmental impacts through producers and consumers. *Science* 360(6392), 987–992. DOI: 10.1126/science.aaq0216.

Ritchie, H. (2021) If the world adopted a plant-based diet, we would reduce global agricultural land use from 4 to 1 billion hectares. *Our World in Data.* Available at: https://ourworldindata.org/land-use-diets (accessed 2 June 2025).

Rogers, S. (2013) *Facts Are Sacred*. The Guardian. [ebook]

UK government (2024) National Food Strategy. https://www.gov.uk/government/publications/a-uk-government-food-strategy-for-england (accessed 2 September 2025).

UN (2024) *Revision of World Population Prospects*. Population division of the department of economic and social affairs of the United Nations secretariat, New York.

Vanham, D., Mekonnen, M.M. and Hoekstra, A.Y. (2013) The water footprint of the EU for different diets. *Ecological Indicators* 32, 1–8. DOI: 10.1016/j.ecolind.2013.02.020.

Wang, T., Masedunskas, A., Willett, W. and Fontana, L. (2023) Vegetarian and vegan diets: Benefits and drawbacks. *European Heart Journal* 44. DOI: 10.1093/eurheartj/ehad436.

Willett, W.C. (1994) Diet and health: What should we eat? *Science* 264(5158), 532–537. DOI: 10.1126/science.8160011.

Zucchinelli, M., Spinelli, R., Corrado, S. and Lamastra, L. (2021) Evaluation of the influence on water consumption and water scarcity of different healthy diet scenarios. *Journal of Environmental Management* 291, 112687. DOI: 10.1016/j.jenvman.2021.112687.

9 Soil as a Living Entity

'Land, then, is not merely soil; it is a fountain of energy flowing through a circuit of soils, plants, and animals.'
— Aldo Leopold

'We are part of the earth and it is part of us...Whatever befalls the earth, befalls the sons of the earth.'
— Chief Seattle (in Kaiser, 1987)

'If the soil is destroyed, then our liberty of action and choice are gone ...'
— W.C. Lowdermilk (1953)

'Upon this handful of soil our survival depends. Husband it and it will grow our food, our fuel, and our shelter.'
— Bhumi Suktam, Atharva Veda

Abstract

Rattan Lal developed a keen interest in soil functions from a young age because soil is the source of numerous essential commodities and thus is rightfully termed 'Mother Earth', and there is a strong interconnectivity between soil health and human health. Thus, sustainable management of soil health is critical to planetary and human health. Adopting conservation agriculture and integrating crops with trees and livestock are useful strategies to improve soil health. The 'One-health' approach is the key strategy to promote regenerative agriculture based on agroecology. Rights-of-soil and rights-of-nature necessitate implementation of the Soil Health Act.

9.1 Soil Health

Humanity's interest in soil goes back to the dawn of time. Many cultures believed that soil is the ultimate mother, which is a source of all the critical substances that humans need. Thus, capacity of soil to provide these substances, or goods and services, has been of interest to humanity since the time immemorial.

In the context of soil's capacity to provide these ecosystem services (ESs), the term 'soil quality' has several components (Fig. 9.1): physical, chemical, biological and ecological. Even from his student days, Rattan Lal realized that most studies have focused either on soil formation and classification or on management of soil's chemical attributes, specifically plant nutrients and factors affecting them. In general, there has always been less interest in soil's physical and ecological properties and processes. Yet, these attributes are critically important to life-support functions of soil.

9.2 The Soil–Life Nexus

Soil and life are intricately interconnected. As Charles Kellogg (1938) of USDA stated, 'Essentially all life depends on soil. There is no life without soil and no soil without life; they have evolved together.' Similarly, Aldo Leopold (1949) stated that, 'Land, then, is not merely soil; it is a fountain of energy flowing through a circuit of soils, plants, and animals'. Roy Simonson (1957) remarked that 'the soil is the link between rock core of the earth and the

DOI: 10.1079/9781836990154.0009

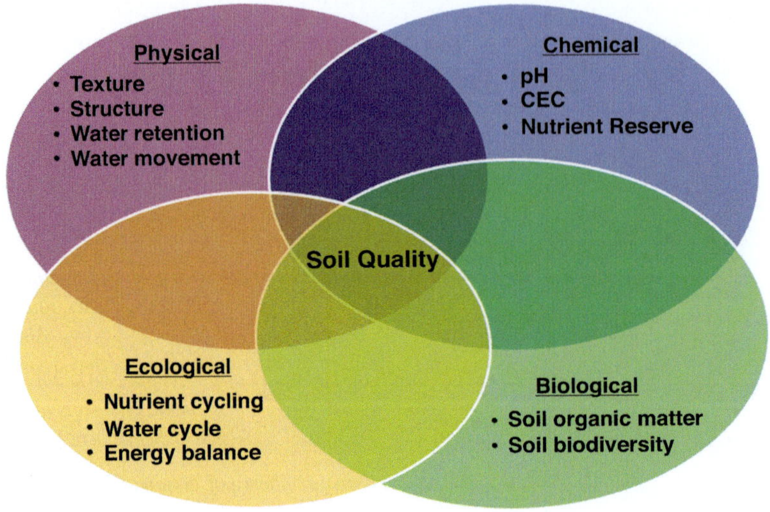

Fig. 9.1. Four components of soil quality are physical, chemical, biological and ecological. These components strongly interact among themselves, and the interconnectivity has a mutually enhancing effect on soil's capacity to support life.

living things on its surface. It is the foothold of the plants we grow. Therein lies the main reason for our interest in soils.' The last statement may be debatable because plants can be grown without soil. Indeed, there is a growing interest in soil-less farming in urban regions using aquaponics, hydroponics, aeroponics etc. Using soil only as a medium for plant growth, as stated by Simonson, may have been the primary cause of soil degradation and desertification. As stated by Marbut (1920), 'probably more harm has been done to the science by the almost universal attempts to look upon the soil merely as a producer of crops rather than as a natural body worthy, in and for itself, of all the study that can be devoted to it, than most men realize'. Yes, indeed, humanity must look at soil beyond the medium to produce 4 Fs: food, feed, fibre and fuel. It is a natural entity, ever-changing and highly dynamic, teeming in and interacting with life, while affecting and being affected by all biotic and abiotic components surrounding it. More importantly, now, during the 2020s than ever before, soil is the largest repository of carbon and moderator of Earth's temperature. Rather than being the source of carbon to the atmosphere, it is and will be used as a major sink of both soil organic carbon (SOC) and soil inorganic carbon (SIC). Thus its management and land use can make soil a part of the solution for adaptation to and mitigation of anthropogenic global warming. In the near future, soil will be a source of several new ESs for humanity and nature. By 2050 and beyond, food production may be a minor role of soil in comparison with numerous known and yet unknown ecosystem services beyond the 4 Fs. Man, or humans, will indeed find and explore new ESs that may bring about a radical transformation in the uses of soil in the near future and beyond.

It is precisely in this context that Daniel Hillel (1981) stated that 'the Latin name for man, homo, is derived from humus, the stuff of life in the soil'. With regards to the importance of humus, Miller and Turk (1943) stated that 'Soils are developed; they are not merely an accumulation of debris resulting from decay of rock and organic materials ... In other words, a soil is an entity – an object in nature which has characteristics that distinguishes it from all other objects in nature.'

9.3 Soil Is a Living Entity

Therefore, the pertinent question about soil is whether this object in nature is a living entity.

Rattan Lal's personal belief, similar to that of many ancient cultures, is that soil is a living entity. But, keeping the personal belief aside, it is important to rationalize if soil is a living entity or not.

One way to answer that question is on the recent (*c*.2010 and beyond) use of the term 'health' by soil scientists, which is normally used only for living organisms. Thus, the use of 'soil health' implies that it is a living entity in nature. However, the interpretation that 'soil is a living entity' is being actively debated by the scientific community. Is soil a living organism, is it a habitat for living organisms, is it merely a natural system that supports life by provisioning of essential ecosystem services, or is it a living organism by itself?

James Lovelock's (1972) Gaia hypothesis suggests that Earth's biological systems and the planet itself behave as a single entity with negative feedback loops that maintain the environments favourable for life. The hypothesis argues that living and non-living parts of the planet interact to create a self-regulating system that maintains favourable conditions for life (Lovelock and Margulis, 1974). If the Earth is indeed a living organism, soil (surface of the planet to around 1 m depth) is the integral component of the living planet.

Soil's capacity to support all terrestrial life depends on interaction between solids (inorganic and organic components), water and air. The interaction between soil (solid) and water leads to water retention and movement, and that between soil and air, or between climate and gaseous exchange, and interaction among all, leads to creation of an environment that supports life (Fig. 9.2).

This life-enhancing interaction among three components (soil/solid, water and air/climate) is highly sensitive to anthropogenic perturbations such as land use and soil/crop/livestock/tree management. Therefore, the choice of judicious land use and prudent agronomic management are critical to protection, enhancement, restoration and sustainable management of the critical life support functions of soil.

As was stated in Chapter 6, there are many known and yet unknown functions of soil which are beyond those of the predominantly agricultural use that has happened since the dawn of settled agriculture over ten millennia to the mid-21st century (*c*.2050). Soil uses and their management will change drastically, to non-agricultural uses, from *c*.2050 and beyond.

9.4 The One-health Concept

The One-health concept states that 'health of soil, plants, animals, people, ecosystems and planetary processes are one and indivisible' (Kemper *et al*., 2020). The importance of soil health to human health was recognized by Sir Albert Howard (Howard, 1943) and Lady Eve Balfur (Balfur, 1948). Several of the essential nutrients (macro and micro) in human diets come from soil on which plants and animals are raised and food is derived from them. Human health is also affected by the environments in which humans live. Three interacting components of the environment are soil, water and air (Fig. 9.2). Land use, and soil/crop/animal management practices, strongly impact the environment quality with the attendant impact on human health.

9.5 Managing Soil as a Living Entity

Soil being a living entity, how should it be managed so that its life-support processes are enhanced and sustained forever, while its basic rights are respected? This is where use of the term 'soil health' is pertinent. Thus, sustainable soil management would imply maintaining its life support process at the desired rate or level forever. Yet, with human-centric focus, management of soil health has been considered with reference to specific functions for the desired land use (agronomic, forestry, pastoral, urban, recreational, mining etc.). With nature or soil-centric focus, however, soil physical properties and processes are critical to maintaining its life support functions. These would include texture, structure, water retention and movement, soil temperature, gaseous exchange, aeration, and depth. Soil physical

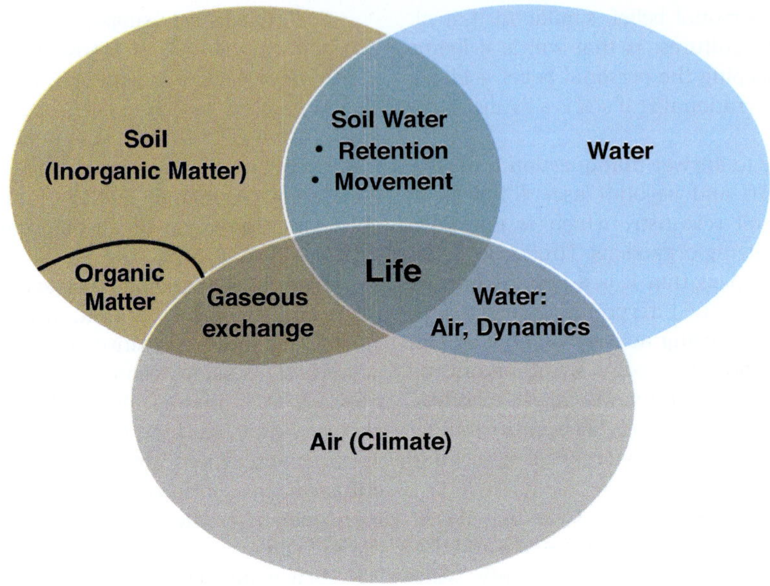

Fig. 9.2. The soil–water–air nexus that forms the basic support system for all terrestrial life on Earth. The interaction between solid (soil inorganic and organic fraction) with water leads to soil water retention and transmission and the so-called green water supply for all terrestrial life. That between water and air leads to precipitation and evaporation as components of the hydrological cycle; and between solid and air to gaseous exchange and moderation of the atmosphere. The Gaia hypothesis is the basis of the central component (life) involving interaction among all three that support life.

properties and functions also moderate its chemical properties and functions (i.e. pH, cation exchange capacity, elemental balance, nutrient retention and transformation), the activity and species diversity of biota (macro, meso and micro), and ultimately soil organic matter (SOM) content, which is the source of energy for all living organisms. Thus, the key management strategy from a soil-centric perspective is to protect, restore and manage SOM content as the source of energy for all organisms in the soil.

Crop residues, and their retention to the soil surface without physical disturbance, is critical to maintaining soil biotic activity species diversity. Therefore, crop residues (both above- and below-ground) must be returned to the soil as food for organisms. This is why it is important for land managers to know that SOM is the heart of soil health (see Fig. 5.6) and soil biota is the bio-engine of the soil as a living entity.

9.6 Resource-poor Small Landholder Farmers as a Mirror Image of the Health of Soil on which They Live

Growing up on a small farm in a rural community during the 1950s, from age 5 to 15, Rattan Lal experienced first-hand the desperation of the smallholder farmers who took away all biomass from the soil: grains and tubers for their food, crop residues for feeding cattle, biomass for fuel for cooking and other uses, and even topsoil for making bricks to construct their houses. There were no fertilizers and the animal dung was also used for cooking and other uses. Thus, soil was continuously being depleted of its SOM stock and plant nutrients. In soils irrigated with the canal water, the only addition of nutrients was from the silt being brought in with the water from the hills and the droppings of some animals when grazed after the crop harvest, and that was not enough. Thus, soils of his rural farming

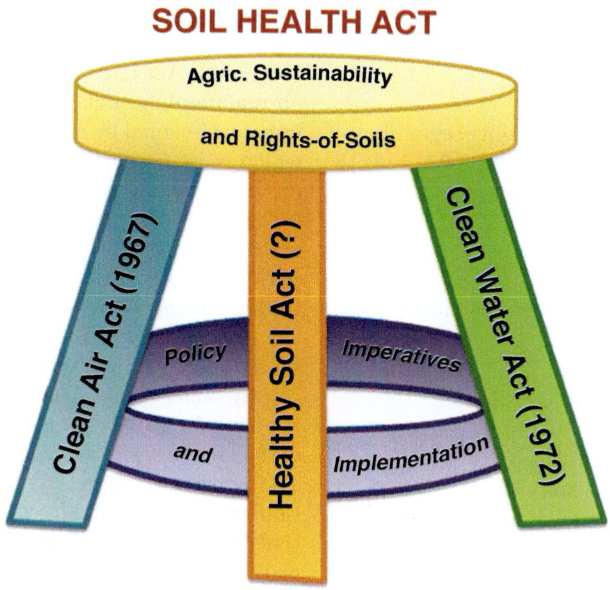

Fig. 9.3. A Soil Health Act is needed to complement the U.S. EPA Clean Air Act and Clean Water Act.

community were on a downward spiral, and the amount and quality of food being produced were neither adequate for good health and well-being of people nor for maintaining the environment. Rattan Lal also realized that people are a mirror image of the soil on which they live. Soil responds to its management. If the management is poor (human-centric), production is low and of an inferior quality, and health of the farm family is poor and on a downward spiral.

9.7 Managing Soil and Life

Because of its importance to all terrestrial life, soil must be managed so that its properties and functions support life and nature's functions. It is this nature-centric approach which has highlighted ideas such as regenerative agriculture (RA) and agroecology (AE). RA and AE are gaining momentum since 2020 in support of the ideology that soil is a living entity. RA is a philosophy, strategy, approach or concept. It does not refer to a practice because there is no 'one-size-fits-all' situation. It is precisely in this context that RA is inspired by eco-innovation, powered by non-carbon energy, driven by a circular economy and green infrastructure, and supported by the recarbonization of the terrestrial biosphere as the bedrock of sustainable development. SOM content, essential to soil health, is the elixir of life and heart of soil health.

Similar to RA there is also a growing interest in AE. It refers to the application of ecological principles to agricultural systems and practices, or the branch of science concerned with this. Indeed, these new farming methods, such as those based on AE and RA, are focused on optimizing (rather than maximizing) agronomic yields while reducing environmental impacts and restoring soil health by improving SOM content and enhancing activity and species diversity of soil biota.

Therefore, RA and AE are strategies of nature-positive agriculture. Basic objectives of nature-positive agriculture are:

- protecting, managing and restoring soil;
- strengthening biodiversity;
- improving water quality and renewability;
- adopting negative-emission farming;
- reducing input of chemicals;
- using biostimulants and organic amendments;

- promoting a biocircular economy;
- supporting the One-health concept; and
- using digital innovations (e.g. precision agriculture).

Some examples of nature-positive agricultural practices based on the concepts of RA and AE believed and practiced by Rattan Lal are:

- conservation agriculture;
- agroforestry;
- integration of crops with trees and livestock;
- positive soil/eco practices;
- CNPK through ISFM
- balanced use of plant nutrients;
- drip fertigation;
- digital agriculture;
- precision farming; and
- artificial intelligence, drones and robotics.

Site- or soil-specific choice of these practices may require modification or adaptation to suit the specific soil/ecological conditions. Regardless of the life-support systems, soil health must be protected, enhanced and sustained through a Soil Health Act (Fig. 9.3). Existing laws of EPA on maintaining quality of water and air must be complemented by enacting a Soil Health Act which must support farmers and land managers with payments for ecosystem services. It is the 'trinity' of these laws (Fig. 9.3) that can make agriculture a solution to restoration of soil quality and improvement of environments through adoption of RA and AE practices.

Rattan Lal argued that there is a strong need to enact and implement a Soil Health Act at county, state, national and international levels. Payment for farmers for ecosystem services (e.g. carbon sequestration) can be a clause of the act. The latter should be relevant to the implementation of the One-health concept which states that the health of soil, plants, animals, people, ecosystems and planetary processes is one and indivisible.

Bibliography

Balfur, E.B. (1948) *The Living Soil*. Faber and Faber Ltd, London.

Hillel, D. (1980) *Fundamentals of Soil Physics*. Academic Press. https://doi.org/10.1016/C2009-0-03109-2.

Howard, A. (1943) *An Agricultural Testament*. Oxford University Press, London.

Kaiser, R. (1987) Chief seattle's speech(es): American origins and European reception. In: Swann, B. and Krupat, A. (eds) *Recovering the Word: Essays on Native American Literature*. University of California Press, Berkeley, California, pp. 525–530.

Kellogg, C. (1938) *USDA Year Book of Agriculture*.

Kemper, K. and Lal, R. (2017) Pay dirt: Human health depends on soil health. *Complementary Therapy in Medicine* 32, A1–A2.

Kemper, K., Xia, Y., Lakritz, J. and Lal, R. (2020) Health of soil, plants, animals and people. In: Lal, R. (ed.) *The Soil-Human Health-Nexus: Advances in Soil Science*. CRC Press, Boca Raton, Florida, pp. 13–45.

Lal, R. (2017) Improving soil health and human protein nutrition by pulse-based cropping systems. *Advances in Agronomy* 145, 167–204.

Lal, R. (2019) Rights of soil. *Journal of Soil Water Conservation* 74(4), 81A–86A.

Leopold, A. (1949) The land ethic. In: *A Sand County Almanac: And Sketches Here and There*. Oxford University Press, pp. 201–226.

Lovelock, J. (1972) Gaia as seen through atmosphere. *Atmospheric Environment* 6(8), 579–580.

Lovelock, J. and Margulis, L. (1974) Atmospheric homeostasis by and for the biosphere: The Gaia hypothesis. *Tellus* 26(1–2), 2–10. DOI: 10.3402/tellusa.v2611-2.9631.

Lowdermilk, W.C. (1953) *Conquest of the Land through 7,000 Years*. U.S. Department of Agriculture, Soil Conservation Service.

Marbut, C.F. (1920) The Great Soil Groups of the World and Their Development. 1927. Oxford University.

Miller, C.E. and Turk, L.M. (1943) John Wiley and Sons, Inc., New York. *Soil Science* 56(1), 67.

Perry, T. (1972) *Home* [film]. Southern Baptist Radio and Television Commission.

Simonson, R. (1957) Soil is the foothold for the plants we grow. USDA Yearbook of Agriculture. https://nmwrri.nmsu.edu/footer_pages/nm-wrri-library-database-files/wrri-library-pdfs/wrrilibrary2/002456.pdf.

Whitney, W.D. and Lanman, C.R. (1962) *Atharvá-Veda-Saṁhitā*. Motilal Banarsidass, Mumbai.

10 Soil Science and Global Issues

'The soil is the great connector of lives, the source and destination of all. It is the healer and restorer and resurrector, by which disease passes into health, age into youth, death into life. Without proper care for it we can have no community, because without proper care for it we can have no life.'
— *Wendell Berry, The Unsettling of America: Culture and Agriculture*

'We must come to understand our past, our history, in terms of the soil and water and forests and grasses that have made it what it is.'
— *William Vogt, Road to Survival*

Abstract

The world is at an ecological and political crossroads. During the 2020s, it was gripped by the 3 Cs: COVID, conflict and climate change. The interaction of the 3 Cs has aggravated several issues facing humanity and the Earth. There are ten global issues (see Preface) and these are driven by the growing and increasingly affluent human population. Almost 50% of the ice-free and desert-free land is being used for agriculture. Therefore, Rattan Lal got interested in sustainable management of soil with his very first professional job as a soil physicist at the International Institute of Tropical Agriculture (IITA). This position offered him the opportunity to work across the humid, sub-humid and semi-arid tropics of the world. This position provided him with the platform to use his ideas on sustainable management of soils, and the implication of land misuse and soil mismanagement to degradation of natural resources.

10.1 Position at IITA

During the early 1970s, when only four international agricultural research centres existed (IRRI, CIMMYT, CIAT and IITA), administrators and scientists from these institutions were appointed by the Ford and Rockefeller Foundations with the support of organizations such as USAID, Canadian CIDA, British ODA, French IRAT/ORSTOM, Belgian AGCD and German GTZ. The CGIAR system had not yet been created (see Chapter 7). The Directors General of these centres (such as Dr Herb Albrecht, Dr William Gamble, Dr Larry Stifel of IITA, Dr Robert Chandler of IRRI, and Robert Havener of CIMMYT) were visionary leaders with big ideas. Their goal was to appoint young scientists from around the world and encourage them to do the best they could in their profession and leave the financial issues to the management. Indeed, this was a golden era in international agricultural research, which created geniuses of Dr Norman Borlaug's stature, the 1970 Nobel Laureate who was then the wheat breeder at CIMMYT.

The offer of a position at IITA was a unique opportunity for Rattan Lal. Rather than management of soil fertility (N, P, K etc.), Rattan focused on the relatively unknown subject of soil physical properties including soil structure (compaction, crusting, erodibility, water retention and transmission), soil temperature regime, drought stress, root growth, mulch farming and soil erosion. The management advised him to establish some world-class facilities so that scientists from sub-Saharan Africa (SSA) and those from the global tropics did not have to go to Europe or North America but could get even better training at IITA. With this mandate, Rattan developed the field and laboratory facilities which were as good as, or

 Making of a Soil Scientist: From a Subsistence Farm to The Global Forum (R. Lal)
DOI: 10.1079/9781836990154.0010

better than, anywhere else. More importantly, during the early stages of his career, Rattan Lal did not have to worry about the funding support which at that time was assured for conducting both basic and applied soil science research. This was the era when the focus shifted from a seed-centric approach in international agriculture to soil-centric strategy.

Rattan Lal focused on studying processes of soil degradation (erosion, crusting, compaction and depletion of soil organic matter content) and assessing the effects of no-till farming as an effective soil and water conservation measure for structurally fragile soils of West Africa (Greenland, 1975; Lal, 1987). Rattan Lal established long-term no-till experiments using run-off plots, field plots and watershed experiments. Indeed, working at IITA from December 1969 to July 1987 was a great research experience for Rattan Lal because IITA provided an important platform for him to address global issues. This position was the foundation for the development of his soil-centric strategy for sustainable management of soil and other natural resources.

He studied the problem of erosion-induced degradation beginning with deforestation by different methods, effects of different methods of seedbed preparation, crop rotation, simple vs complex farming systems, and in-field measurements of soil physical and hydrological properties.

10.2 Watershed Management to Enhance and Sustain Soil Health

During the mid-1970s, when Dr William Gamble was the Director General and Dr Dennis Greenland (an eminent soil scientist (see Greenland, 1975)) was the Director of Research, Rattan Lal established a watershed management experiment at two locations (Ibadan and Benin) to assess the impacts of methods of deforestation, land use and soil management on the hydrological cycle, micro- and meso-climate, soil erosion and productivity. The estimated budget was about US$ 1 million over a ten-year period with about half of this for the initial year of installation of the equipment for monitoring and measurement of soil, climate, environment and agronomic parameters. Dr Gamble travelled to New York to obtain the funding support with a lot of expectation. However, he telephoned from New York to inform Rattan Lal that while financial support was made available, he refused to accept it. Confused and disappointed, Rattan Lal asked Dr Gamble why this was. Dr Gamble replied in a matter-of-fact manner: 'Because some money is poisonous'. Rattan Lal has never forgotten that statement. Nonetheless, Dr Gamble assured Rattan Lal to continue the good work and he would find another source which is good for IITA's long-term mission. Such big-hearted and visionary research and academic managers are rare and a matter of great fortune to work with.

During his PhD studies at the Ohio State University, Rattan Lal had visited the North Applachian Experimental Watershed (USDA-ARS) station at Coshocton, Ohio. Practicals on groundwater pumping tests in the hydrogeological class (1967) were done at Coshocton. From his trip to Coshocton, he was familiar with the weighing monolith lysimeters, H flume, Coshocton wheel sampler, broad-crested weirs, and other hydrological facilities. Dr Gamble had asked Rattan Lal to find out if any scientist from Coshocton could come to IITA and advise in establishing the watershed management experiments. Rattan Lal suggested that the Director Emeritus of the Coshocton Research Center (Dr Lloyd Harrold) should be invited. Dr Gamble made arrangements and Dr Harrold came to IITA for three months each year for 3 years. This was a most useful co-operation and a professionally rewarding experience for Rattan Lal.

Another notable incidence of administrative support was the need for a shearing blade to cut trees at ground level but without digging out the roots and scraping the topsoil. The dozer-mounted shearing blade is heavy (weighing 1–2 tons) and was not available in Nigeria. The shearing blade was airfreighted from Rome and delivered to IITA in about 2 weeks.

10.3 Globalization of Soil Management Issues

Now, Rattan Lal was an independent scientist with experiences of working in India, USA,

Australia and Africa. From being a mentee, he was now a mentor in soil science, and the research done at IITA planted the seed of the 'soil-centric green revolution'. While at IITA he advised on the thesis research of about 40 scholars from around the global tropics.

10.4 Importance of Visionary Leaders and Dedicated Research Managers

When IITA laboratory buildings were under construction, the Board of Trustees advised designing the buildings to have a long life and need little external and internal maintenance. Making investment in the beginning is much better than having frequent maintenance. These were the views of the trustees of IITA from Rockefeller and Ford Foundation, USAID, ODA, CIDA, IRAT, ORSTOM, ACGD, JAICA etc.

Yes, these were the visionary leaders with big ideas and bigger heart and they knew of donors who had even bigger ideas and deeper pockets. This ideal combination made the 1960s through 1980s a golden era of international agricultural research, and Rattan Lal benefitted from this age of doing something good for the global tropics.

Now that the world is facing climate change during the early 21st century (*c.*2000–2030) and other environmental issues, there is even a stronger need for leadership and donor organizations that can effectively address these issues: leaders who translate their words and slogan into actions, donors who are committed to implementing game-changing solutions, and scientists who can develop a credible database on techniques of land-based solutions which are pro nature, pro agriculture and pro farmers. Now is the time to make agriculture a part of the solution. Rather than focusing on the maximum agronomic yield, the goal must be to produce more from less, restore soil health, harness the potential of land-based sinks for atmospheric CO_2, adopt nutrition-sensitive agriculture, promote carbon farming or commodification of terrestrial carbon stock which can create another income stream for farmers, and adopt a Soil Health Act at county, state, nation and global scale. More importantly, this is the time to translate political rhetoric into action and enact policies that reward and motivate farmers, ranchers and land managers to adopt practices based on the concept of regenerative agriculture and agroecology (see Chapter 9).

10.5 Major Research Findings of Rattan Lal from IITA

Major highlights of Rattan Lal's research at IITA from 1969 to 1987 can be summarized as follows (Lal, 1987): (i) if new land clearing must be done for horizontal expansion of agriculture, it should be done manually or with shear-blade equipment so that detritus material and tree roots are left on/in the ground; (ii) there should be no in-field burning of fell biomass and it should be allowed to decompose *in situ* or in windrows; (iii) a rapidly growing and an aggressive cover crop (e.g. mucuna, stylo, centrosema, kudzu, guinea grass, switch grass) should be established immediately after clearing to provide a soil cover; (iv) grain/food crops should be grown through the residues of suppressed cover crop by using a no-till seeder and following the basic principles of conservation agriculture; (v) crop residues must be left on the ground as surface mulch and neither removed nor burnt; (vi) supply of plant nutrients must be maintained through integrated soil fertility management (ISFM); (vii) crops must be integrated with trees (as alley cropping by establishing perennials on the contour about 4m apart), with livestock based on controlled grazing; (viii) use of heavy machinery must be minimized to reduce risks of soil compaction and crusting; and (ix) activity and species diversity of soil biota (earthworms, termites and meso- and micro-organisms) must be maintained at a level of about 5 Mg of live biomass per ha because biota is the engine of soil health and its sustainable management. In other words, the strategy is keeping the soil always covered with crop residue mulch or a cover crop, growing deep-rooted legumes (pigeon pea, stylosanthes, mucuna) in the rotation cycle, and growing food crops (cereals, legumes, root crops) by conservation agriculture through integration with trees and livestock.

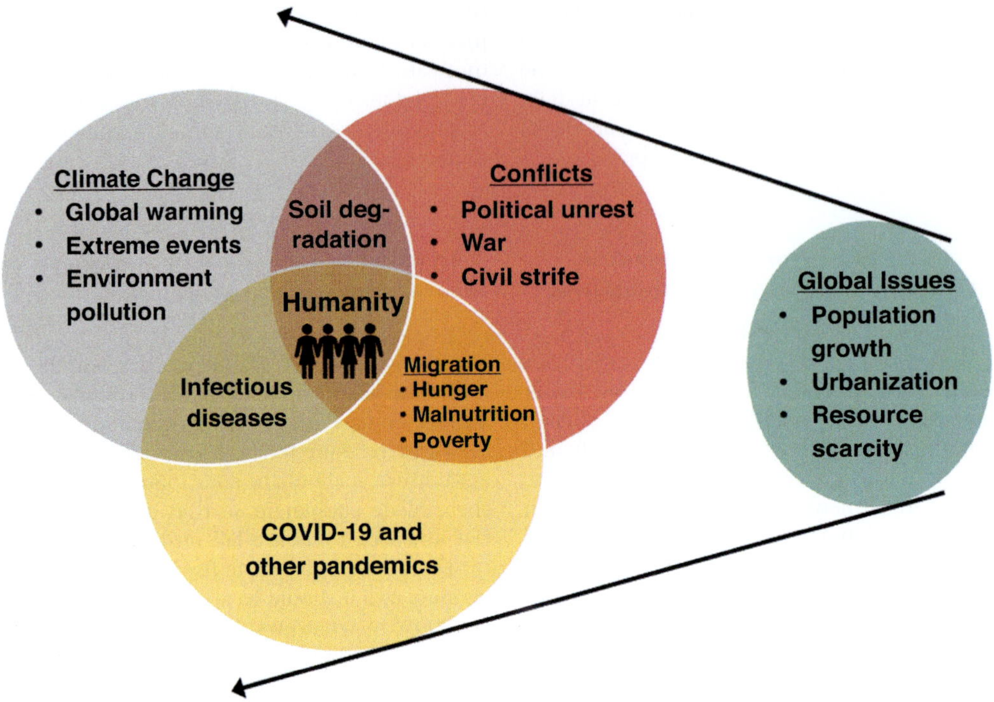

Fig. 10.1. The vicious grip of the 3 Cs on humanity in the 2010s and 2020s.

10.6 Managing Soil to Address Global Issues of the 21st Century

The world is at an ecological and political crossroads. It is also gripped in the 3 Cs: COVID, conflict and climate change (Fig. 10.1). The interaction of the 3 Cs has aggravated the problems of soil and environmental degradation, accelerated global warming, aggravated the risks of civil unrest and political instability, and led to land grab and wars.

10.7 Soil and International Art and Culture

Because of large variations in soil colours (hue and chrome), soil is widely being used in preparing portraits and cultural drawings. Indeed, soil is an important teaching media. Some examples of using soil in paintings are shown in Fig. 10.2, the logo of the Ohio State University, and Figs 10.3 and 10.4, as the title of the Lal Carbon Center. These paintings were prepared by Alexander Jones (Fig. 10.2) and the visiting scholar Dr Toru Nakajima in 2024 (Figs 10.3 and 10.4). The sign 'Lal Carbon Center' was prepared by using seven major soil series recorded in Ohio (Fig. 10.4).

A soil scientist from India, Dr Gopal Mahajan, is very professional in making portraits from diverse soils. Soil used in Fig. 10.5a is from Goa, and has a deep red colour caused by high content of iron at 120–200 cm depth. The black portrait (Fig. 10.5b) is made from black soils of Indus from the Jalgaon district of Maharashtra and the portrait depicts the unique cracked characteristics of swell-shrink of these soils. Dr Mahjan believes that, in addition to being an innovative strategy to promote soil science education, this soil-based art is a down-to-earth approach to connect people with soils and bring families together.

There are many uses of soils beyond production of biomass for food, feed and fuel. It is the major reservoir of germplasm, principal source of freshwater and its renewal and purification, repository of the terrestrial C stock, and moderator of climate. Soil microbiota are the primary

Fig. 10.2. The Ohio State University logo soil painting. (Photo: Mr Alexander Jones for an exhibit for the Ohio State Farm Science Review 2024)

Fig. 10.3. The Lal Carbon Center title painting prepared by Dr Toru Nakajima for an exhibit for the Ohio State Farm Science Review 2024.

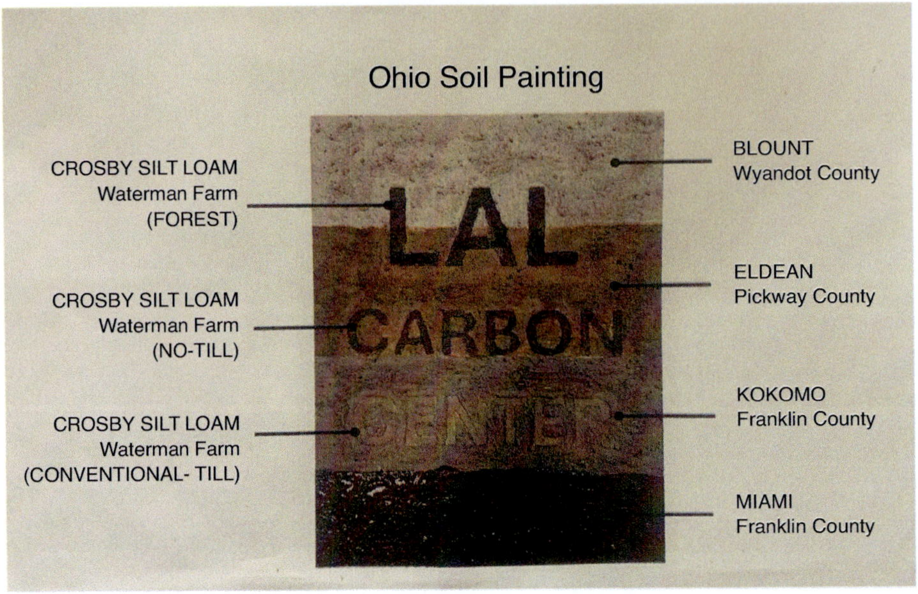

Fig. 10.4. The Lal Carbon Center title painting prepared by Dr Toru Nakajima showing the soil series used for each colour.

source of antibiotics used to treat diseases of humans and livestock. Soil is a source of raw material for industry (beyond pottery). Soil is an archive of human and planetary history.

Soil also has numerous cultural, spiritual and aesthetic values. It is a source of beauty and has been used in art for a long time. With a wide range of colours, soil is being used as a component of art science and a technique to promote soil science education (Fig. 10.2).

10.8 World Peace and Soil Health

Famine, migration, poverty and wars are human-made tragedies. The long-lasting solution to these tragedies is a scientific issue involving protection, restoration and sustainable management of soil health through adoption of RA and AE. Global political leaders must give peace a chance by considering it a scientific issue related to sustainable management of finite, fragile and precious natural resources such as soil, water, air, landscape and biodiversity.

Most of these and other problems are driven by population growth, urbanization and related issues. However, sustainable management of soil and agriculture cuts across most of these problems and other emerging issues. Yet, the human demands must be reduced by controlling population growth so that the world population may peak by 2050 or sooner and begin to decline thereafter. That being the case, there should be an agreed timetable to return some land back to nature (see Table 8.4).

Yes, world peace is a scientific issue, and management of soil health is critical to achieving global peace and stability.

Bibliography

Berry, W. (1978) *The Unsettling of America: Culture & Agriculture*. Sierra Club Books.

Greenland, D.J. (1975) Bringing the Green Revolution to the shifting cultivator. *Science* 190, 841–844.

Lal, R. (1987) Managing soils of sub-Saharan Africa. *Science* 236, 1069–1076.

Vogt, W. and Warde, P. (2013) Road to survival (1948). In: Robin, L. and Sörlin, S. (eds) *The Future of Nature: Documents of Global Change*. Yale University Press, New Haven, Connecticut, pp. 187–194. DOI: 10.12987/9780300188479-018.

11 The Lal Family

'Grandma, I want to talk with grandpa. No, you can't, because he is asleep. Did he eat his dinner. No, he does not eat dinner. Grandma, please tell grandpa that if he does not eat dinner, he will not grow up.'
— Jay Henry Bhakri Lal (age 5) talking with his grandmother (Sukhvarsha Lal) on 9 March 2025

'Family is not an important thing, it is everything.'
— Michael J. Fox

Abstract
Despite his education in the USA and Australia, and accepting a position with the International Institute of Tropical Agriculture in Nigeria, Rattan Lal was deeply committed to his culture and traditions. Thus, he returned to his family and the brathery to follow the culture of starting a family through an arranged system of marriage. He followed the advice of his father who stated that getting education is one part of life but getting married and starting a family is another. Keeping these two parts separate and yet interactive is the pathway to a successful professional and family life. Before raising his own family, Rattan Lal had provided funding support to his father and brother to build a new house and purchase some farmland equivalent to, or more than, what he had in Pakistan. However, Rattan Lal made a socio-cultural mistake of disconnecting himself from his family and rural community. This disconnection had many important implications.

11.1 Transition from Student to Professional Soil Scientist

Rattan Lal began his new appointment as a soil physicist at the International Institute of Tropical Agriculture (IITA), Ibadan, Nigeria, in December 1969 (see Chapter 7). It was a challenging task to get adapted to new job, the tropical climate, and new social, cultural and economic conditions. For the first time ever, Rattan had a car, and it was provided by IITA. The new Ford Zephyr (1969 model) was impressive. Rattan Lal was also assigned a three-bedroom apartment within a complex on the western part of Ibadan and close to the so-called Moore Plantation, which was a government-operated agricultural research institute. In addition to Rattan Lal's, there were two other IITA staff families living in this complex. Thus, Rattan Lal was a proud owner of a new car and had an impressive apartment, and these were pleasant experiences for him. Another new experience was having a house servant. It was customary for expatriates to have a cook/steward to cook food, clean the apartment, wash and iron clothes and do the grocery shopping. The cook/steward was paid about 10–15 pounds per month and he and his family lived in servant's quarters attached to the apartment building. In those days, it was easy to find a cook/steward who knew how to prepare British and/or Indian food. Road traffic in 1970 in Ibadan was rather light. Thus, Rattan Lal and other staff came home for lunch, as 1 hour was adequate to come home (approx. 10–15 miles) and return to IITA for the rest of the day. Having a cook/steward also made it easy to organize social events such as dinner parties for colleagues.

DOI: 10.1079/9781836990154.0011

11.2 Inauguration of IITA

On the very first day of his joining IITA, Rattan Lal was introduced to other scientists who had also joined during 1969, a few months ahead of him. There were three top administrators: Dr Herbert Albrecht, Dr Rodney Briggs and Dr Jim Mitchell. These three were supported by other managerial staff including Dr Arthur Leach, John Craig, Alec Butler, Wolley Shoyinka, and others. Among scientists were two soil scientists (Dr Frank Moorman, and Dr B.T. Kang), five plant scientists (Dr Sadiki Sadik, Dr Arthur Lamey, Dr Ken Rachie and Michael Harrison), three plant protection scientists (Dr Keith Whitney, Dr Rob Williams and Dr Keith Moody), and one economist (Dr Doug Headley). Paul Zuckerman was a research scholar who was conducting his PhD thesis project involving on-farm surveys and addressing social economic issues. Two other plant breeders who joined IITA in 1970/71 were Dr Sangki Haan, a root crop breeder, and Dr A.C. Abifarin, a rice breeder with focus on upland rice.

IITA was officially inaugurated on 21 April 1970, about four months after Rattan Lal had joined the institution. The guest of honour at the opening ceremony was the then Head of State of Nigeria, General Yakubu Gowon. Two of the six major buildings of IITA were already completed, and these were the farm building and the mechanical shop-cum-repair garage. Other buildings under construction included two research buildings (400 and 401), the administration building (500) and the international house, which comprised dormitories, cafeteria, swimming pool, tennis courts and volley-ball ground. In close proximity to the international house was the golf course. Staff houses, to accommodate all scientists (32 planned at the time) on campus, were also under construction. The inauguration ceremony was held in the impressive and spacious farm building. However, the planned field visit to some of the experiments, established in a hurry, to be shown to the Board of Trustees and other visitors, had to be changed because of a serious rainstorm the evening before and the serious soil erosion it had caused to all the ploughed plots (see Chapter 7).

During the 1970s and 1980s, there were some prominent research scholars who conducted their PhD thesis research. In addition to Paul Zucherman, others included Patrick Aina, in soil science (later Vice-Chancellor of the University of Ife, Nigeria), Mambani Banda, in soil science (later Rector of the National University of Zaire at Yangambi, Zaire), Nteranya Sanginga, in soil science (later Director General of IITA in 2011), and Simeon Euhi, in economics and social sciences (later Director General of IITA in 2023). Indeed, IITA was a magnet for brilliant young scientists who contributed enormously to agricultural research and development.

11.3 Working with the University of Ibadan

The IITA campus was located about five miles north of the well-established University of Ibadan, which at that time was also affiliated with to University of London. The University of Ibadan had a department of agronomy, which consisted of faculty of soil and plant sciences. The head of the agronomy department was Prof. A.A. Fayemi, who was an agronomist/soil scientist. Two other soil scientists were Dr A.A. Agboola and Olaolu Babalola. Babalola registered for the PhD degree at the University of Ibadan and was the first-ever graduate student with Rattan Lal. Babalola completed his PhD thesis research at IITA. Dr Babalola later became head of the department of agronomy of the University of Ibadan. Dr Agboola was a well-known soil fertility management expert. Dr H.R. Chheda was a plant breeder at the University of Ibadan, and he had a close co-operation with the plant breeding programme at IITA.

This community also had a close-knit social group. During the first decade after inauguration (1970–1980), IITA had many international conferences including those focused on soil and water conservation, methods of deforestation, farming systems, crop breeding etc. The Director of Research from 1975 to 1977 was Dr Dennis J. Greenland, who organized some of these events, and Rattan Lal was the principal scientist involved in three of these conferences: in 1975, 1979 and 1982. Most of these conferences had social events organized at the International Center, and scientists form Moore Plantation and the University of Ibadan were also invited to these events.

11.4 New Farmland and Home for Father and Brother in Rajound

After the first visit to his family in Rajound (Haryana, India) in December 1969, when he was en route from Sydney to Lagos with a stopover in New Delhi, Rattan Lal visited his family in December 1970 and also in December 1971. IITA had the provision for its staff to take one-month vacation to the home country with the passage paid for the entire family. When Rattan Lal visited in December 1970, his father had constructed a new home which comprised of two bedrooms, one kitchen and a large sitting- or family room. The house had two distinct parts: presumably one for Rattan Lal and the second for his elder brother. There was a veranda in front of both bedrooms. The house also had a confined courtyard and a hand pump for water, and it was equipped with electricity. This was a great improvement over the traditional one-room family home made from mud bricks where Rattan Lal had spent his childhood. Now he had his own bedroom which was about 15 ft x 15 ft and connected through a door to the sitting-room.

Sleeping for the first time in the new bedroom during December 1970, Rattan Lal was reminded about the stories narrated by his bebe. She always cautioned that when you put on your *jootty* (traditional footwear made from raw cowhide by a village shoemaker and left outside of the house for the night) in the morning, it is a good idea to shake it with a stick to be sure that there are no snakes or scorpions hiding in it. Snakes are specifically looking for a warm place to hide during the winter months, and hiding under a warm quilt can be a favourable choice. As Rattan Lal was getting ready to sleep, he felt something crawling beneath the quilt and around his feet. He got out of the bed in a hurry and called his brother. His brother, carrying a bamboo stick, came and lifted the quilt. Sure enough, there was a snake. Rattan Lal and the family (young children living at IITA) had similar experiences with snakes in their home and in their garden in Nigeria. Encountering snakes in experimental plots at IITA, especially under the dense canopy of a cover crop such as that of *Mucuna utilis*, was not uncommon and wearing long boots was advisable. Not only does soil under a thick canopy of a cover crop have lower temperature, by 10–15°C, than that under bare soil surface, but frogs and other fauna under cover crops are also an attraction to snakes. Rattan Lal's experience with snakes at the village farm continued throughout his stay at IITA from 1969 to 1987.

During his visit in December 1970, when the new house was already constructed, Rattan Lal also provided additional money to his father to purchase some farmland equivalent to or more than what he had left behind in Pakistan. Thus, during his visit to the family in December 1971, his father had the proud honour of both a new house and a respectable farm, and Rattan Lal was very happy to see that. In addition to the family-owned 1.5 acres, his father purchased an additional 12 acres of new land. During 1970/71, farmland in Rajound was available at about Rs 4000 per acre (at the time, equivalent to US$800). Thus, 12 acres was worth about US$ 10,000. Rattan Lal suggested to his father that this land and house should both be in the name of his elder brother. There were some objections to this suggestion from his elder sister and aunt, but his father and brother were happy with this decision. Rattan Lal wanted to make sure that his elder brother, who was neither educated nor had any other source of income, had his own farm and house to raise his family. Thus, the new farm and new house were owned by his elder brother. In addition to his elder brother's wife and six children (two sons and four daughters), Rattan Lal's father and his uncle (elder brother of his father) also lived in same house. This was a great accomplishment for Rattan Lal.

This decision of not having Rattan Lal's name on the farmland and house, seemingly an admirable cause and with good reason at the time, so that his brother had a house and farm for himself, also created a strong disconnect for Rattan Lal with his village and the community, or brathery. After his father and uncle died in 1978 and 1979, respectively, Rattan Lal lost contact with his family and the village community. With hindsight, not keeping his name on the home and land was a major social/cultural mistake (see Chapter 12). It created a disconnect between Rattan Lal, his village, family and community. In addition, as he learned later to his disappointment, none of the children of his brother completed even a primary school education. Having the land and house gave them a false sense of security so that they felt no need to study. In fact, all of them became over-dependent on Rattan Lal, which was a major tragedy. Thus, giving the home and land away, noble and worthwhile as it then seemed, turned out to be a process of de-rooting of Rattan Lal from the family and community in which he grew up. Despite the obvious disconnect, especially after the death of his father and uncle, and later his brother, the culture and tradition of the community, as narrated to him by his grandmother, never left Rattan. Traditions of following vegetarian food habits, a simple lifestyle, hard work and commitment to the family were all deeply rooted in him and stayed with him his entire life. The ideology was embedded in each cell of his body and was the guiding principle of his professional career and personal life. Yes, the ideology that he learnt from his grandmother never left him during his long journey over six continents and six decades. During his later career, in 2010 and beyond, he wrote several scientific articles which were based on the ancient Vedic culture and Sanskrit literature (see Chapters 1 and 5). Rattan Lal had left India, but India never left him.

11.5 Arranged Marriage

All marriages are arranged. But there are two types of arrangements: one arranged by two families, and the other arranged by two persons getting married. Both types are good systems. In the family-arranged marriage, the love grows after the marriage. In the individual-arranged marriage, the love has reached its peak when the two people get married. In the first case, it is marriage before love, and in the second it is love before marriage. It is a matter of personal choice.

One key example of Rattan Lal following the Indian tradition was that his marriage was arranged by the two families (his and his would-be wife's). After living in the USA, Australia and Africa, Rattan Lal came back to his community to get married following the tradition of his family, community and the village brathery. During 1970 and 1971, while in Nigeria, Rattan Lal met one Indian family who had a marriageable daughter and the family was interested to pursue the matter. However, Rattan Lal was concerned about two issues: one was the highly educated and elite family background, including that of the daughter – they were all well-to-do and professionally settled. Rattan Lal was not sure how this educated family would react to the village-dwelling farmers in a remote area in Haryana, which had neither formal education nor a modern way of life. The second was the caste. Rattan Lal's brathery (community) in the village and his family were rather traditional and belonged to an orthodox Brahmin family, which strictly followed the ancient Vedic culture or *Snatan Dharma*. His father and community always talked about them belonging to a Brahmin culture, and thus Rattan Lal was not sure how his brathery and family would react to a non-Brahmin bride. There were also similar options of marriage among some acquaintances of his class fellows in India, but both concerns listed above kept Rattan Lal away from pursuing those options. His class fellows and friends often backed out stating that the family of the would-be bride was not interested in sending their daughter abroad, and especially to living in Nigeria. Sending the daughter abroad might have been an excuse or a cover-up to hide the real social issues, which were by-and-large ramifications of the two points stated above.

Meanwhile, Rattan Lal was 28 years old (according to his official date of birth but probably 30 in reality). His age was becoming a major concern to his father, bua ji and elder sister. During those times, and in rural communities, boys and girls were married at 18–20 years of age. Thus, according to the standard of the rural community, Rattan Lal was considered 'an old man'. His family was worried that he may never get married, because no one in the community would be willing to marry their daughter to an older person. Furthermore, the girls in his community were mostly primary or secondary school dropouts. There were few if any who had graduated from high school. His father often remarked, 'I did not know that Rattan Lal will keep studying forever, and continue his studies beyond 10th, 14th, 16th, 20th and 21st grade'. He was obviously referring to diplomas form matric, BSc, MSc, PhD degrees and postdoc experience. His brathery would often remark about the karma, 'This is what is in his karma, and that no one can change the 'karmagati'. That is the way it was supposed to be, and nothing can be done about it. *Karmgati tare nahin tare*' ('One's fate can never be changed, it is pre-decided by the divine power, and one must accept it as is'). In this context, A.L. Basham (1954) also observed, 'We cannot escape the law of karma any more than we can escape the law of gravity'.

At the time of publication of this autobiography, Rattan Lal and Sukhvarsha had completed 54 years of their first of seven lives of marriage that the Hindu priest pledged them during the Vedic ceremony. The Swastika, borrowed by some western powers during World War II, is the traditional Vedic religious symbol. The other symbol, which is tattooed on Rattan Lal's right hand since he was about 8 years old, is that of 'Om' or 'Aum' (the primordial sound of the universe and encompassing all creation).

And then, on an auspicious day, a distant acquaintance of the community happened to come and visit Rattan Lal's father. The visitor suggested that 'there is a Brahmin family, also migrant from Pakistan, and it has a college-graduate daughter. Furthermore, the father of the would-be bride is educated in England.' The visitor also reminded Rattan Lal's father, who was already awestruck with whatever he had heard thus far, that 'You may know the father of the girl because he was officer-in-charge of building and maintaining the irrigation canals in the west Punjab region where we all come from. If you like, the acquaintance said, I will talk to the father and show him the photo of your son.' Rattan Lal's father was very enthusiastic about what was being said. He was happy to show the visitor a photo of Rattan Lal and assured him that 'Rattan Lal is indeed very youthful looking and not so old as our community thinks'. However, either Rattan Lal's father forgot to mention, or did not want to say, that he already had some grey hairs. In a few days, the distant acquaintance returned with the good news that the bride's family had agreed to marry its youngest daughter to Rattan Lal, and he even brought with him a photo of the would-be bride. The photo of the bride-to-be was immediately airmailed to Rattan Lal with the happy note that he was going to get married after all.

Rattan Lal wrote back: 'The good news is, I would like to meet the family and personally see the girl during my visit to India in December 1971 and then we will decide'. His father and family replied:

> There is nothing to see, and the matter is already decided. Your sister and sister-in-law went to see the family. They are very happy to state that girl is very healthy, she has a good appetite, and looks to be well-fed as those from well-to-do families are expected to look. She does not look like a shirt put on a hanger. She has wheatish complexion and is of an average height. Because she is a college graduate she will fit very well with your English-speaking colleagues in

Fig. 11.1. Rattan and Sukhvarsha Sharma were married on 12 December 1971. The marriage ceremony was performed by a Hindu priest according to the Sanatan Dharma Vedic traditions in front of a ghee-fueled Havana-fire over the Swastika symbol. (Photo: Rattan Lal).

Nigeria. It is important for you to know that in our community we neither show our girls to anyone nor do we see others' girls. Showing and seeing girls is neither our culture nor a respectful tradition.

Rattan Lal's brother-in-law, husband of his elder sister Sita, also mentioned that he 'did not see his sister before marriage and neither did your brother see his wife or your sister-in-law. So, what are you complaining about? Don't be so arrogant with your higher degrees from the USA or Australia. It is the family and community tradition that are the important consideration.'

When Rattan Lal insisted that it would be nice to see his would-be wife, finally his sister-in-law joined in the discussion. Her name was Sheela and she was a very charming woman. But in front of Rattan Lal's father and uncle, she always kept her face covered and spoke rather softly, as a daughter-in-law was expected to. She assured Rattan that 'his would-be wife is very much suited to him, and it is truly a match made in heaven'. In Punjabi culture, the sister-in-law and the younger brother of the husband have a very special relationship. She can be like a mother, elder sister, friend and a confidante or an advisor all rolled into one person. Rattan Lal's sister-in-law, about ten years older than him, was a beautiful woman. While she did not know how to read or write or even sign her name, she would look very much like a European woman if she were to dress up in western clothes. Finally, she whispered in Rattan Lal's ear, 'Why don't you trust me, naughty boy. Is there something special that you want to see her for? I can do that for you and let you know, but for a modest fee of some bangles or a new shawl.' She pinched him and left, and that personal conversation settled the matter.

Indeed, everything was so simple and straightforward. There was nothing more to be said nor done. The matter was decided, according to the karma, which no one can change, not least the western-educated and so-called English-speaking Rattan Lal. Everything was all fixed by destiny. The family and parents were merely carrying out God's pre-decided wishes and these wishes should be accepted as God's decision. The whole process could not be any simpler. Nobody could understand what all the fuss was about. Both families were busy in getting ready for the marriage, already fixed on the muharat fixed by the priest as 12 December 1971 at 10 am IST. The date and time were fixed when the priest consulted the religious almanac called the patri.

The maiden name of Rattan Lal's wife is Sukhvarsha Sharma. Being from the same caste, Rattan Lal's full name is also Rattan Lal Sharma. However, Rattan Lal's sub-caste (within the Brahmin caste) is Bhakri. Bhakris are Saraswat Brahmins and are spread over widely separated regions spanning from Kashmir and Punjab in north India to Konkan in west India to Karnataka and Kerala in south India. The word Saraswat is derived from the Rigvedic Sarasvati river.

Thus, Rattan Lal's full name should have been Rattan Lal Bhakri. However, his last name was never entered in the register of the village school. Thus, following western customs, Rattan Lal started a new family name, Lal, which is, in fact, his own or given name rather than that of his family. In addition to both families belonging to the Brahmin caste, both were also native of west Punjab, which was by then in Pakistan. Rattan Lal's family was from a village named Karyal in district Gujranwala (Fig. 1.1) and his wife's family came from the village Baddoken Gussian in district Sialkot. The fact that both families were originally from west Punjab was another important deciding factor.

The marriage date (*muharat*) was fixed, by pandits, according to the location of the stars, sun and moon on the date of birth of the bride and groom and according to the first letters of their names. Rattan Lal's exact date of birth was not known, which may have caused some errors in deciding the exact date. Nonetheless, the pandit was sure that the chosen date was 'absolutely superb and all stars were perfectly aligned'. The priest mentioned proudly that this was the most perfect star alliance that he had ever observed in any marriage ceremony he had performed. Obviously, he was hoping for some extra fee from both the families. However, the auspicious date for the marriage, 12 December 1971, was in the middle of the Bangladesh war between Pakistan and India and there was darkness everywhere after the sun set. Home lighting was banned during the war all over northern India. Rattan's family rented a bus and invited about 25 members of the Brahmin brathery to attend the wedding which was held in the city of Karnal, about 100 miles north-west of New Delhi and 40 miles east of the village of Rajound (Fig. 1.1).

Following the family tradition, the wedding dress for the groom (or the bride) was provided by the maternal uncle. Thus, Rattan Lal's dress (pants, shirt, jacket, turban etc., all in white) were to be provided by his maternal uncle or the brother of his deceased mother. Rattan Lal knew well that his maternal uncle, whom he did not know very well except for having seen him during the weddings of his elder sister and brother, did not have enough resources to buy the expensive dresses. Thus, Rattan Lal got his dress tailor-made and gave it to his maternal uncle so that he could present it to him as his present to the groom. The turban was put on the head by Rattan Lal's father. Rattan Lal told his father that he would tie the turban himself. This was not appreciated because, following custom, the turban must be tied by the father. The Brahmin community in Rajound had a common groom crown made of silver and it was loaned to the family for the occasion. The silver crown was also tied to the turban, and a sehra (string of garlands made of silk or artificial flowers) was also tied to the crown and these covered his face, while a priest was reciting the Sanskrit mantras.

Rattan Lal rented four cars to lead the procession, to show that the groom was indeed a well-to-do person. However, Rattan Lal rode on a specially rented mare, from a company that specialized in renting mares for weddings. The mare was specially decorated for the occasion and it came with a trainer who handled it. Rattan Lal rode the mare, the first time he had done so, and his nephew (about 10 years old) sat behind him as his *sarbala* (groom-mate). Rattan Lal's sister held the reins of the mare and she and other women of the family sang some folksongs for the auspicious occasion. Women also braided hairs of the mare. The marriage party, led by a traditional music band (also rented), four rented cars, the groom riding on a mare, and the members of his family and brathery walking behind, arrived at the home of the bride at about 9 am. The band of musicians were playing some romantic tunes from modern Bollywood movies. Neighbours were standing in the street and on rooftops, watching the procession marching towards the home of the bride. The bride and her family were waiting at the door of their home to receive the groom and the marriage party, called *barat*. The would-be bride was being escorted by her mother, sisters and sister-in-law. With the help of the caretaker of the mare, the groom dismounted.

Elderly folk from both families greeted one another in a custom called millni, or greetings. Fathers of the groom and bride, mothers of the groom and bride, elder brothers of the groom and bride, and elder sisters of the groom and bride greeted one another, put garlands in one another's necks, and in this case the relative of the bride gave some gift (a shawl, blanket, sari etc.) to the counterpart of the groom. On behalf of the groom (Rattan Lal's family), the father was represented by his taya (great uncle) and mother by his bua ji (great aunt). This was an appropriate gesture made by the family to respect two elder members of the family.

After the family *millni* (introduction) was done, it was the turn of the bride and groom to greet one another. Rattan Lal was guided by his father and accompanied him to where the bride was waiting along with her father, mother and other members of the family. They both exchanged garlands, and both were guided to a specially prepared site in the courtyard where the priest was preparing the holy fire (*havan*) for the ceremony. Fire worshipping was done in all religious ceremonies according to the tradition of the Sanatan Dharma. All mantras were recited in Sanskrit that no one fully understood, with perhaps the exception of the priest.

Rattan Lal's father-in-law, trained in hydrology from Southampton, UK, wore a typical English-style hat and pantaloons. He asked the priest in a matter-of-fact way, 'How long will the ceremony take?' The priest replied that '*Jajman Ji*, it takes about 3 hours'. The duration of 3 hours was not acceptable to the father of the bride because the marriage party had to return to their village before sunset. Thus, the father-in-law suggested to the priest that 'he will get double the amount of *dakshina* (fee) if he were to finish the ceremony in 1 hour, and no more than half the normal rate if it was to take any longer'. The priest was happy to oblige and finished the ceremony in less than an hour. The sacred fire was started in the front yard of the bridal home, and all members of the bride's and groom's families sat around the sacred fire. Next to the sacred fire was made the symbol of the Swastika, which is the worship sign of Indo-Aryans (unfortunately defamed during World War 2). The priest started chanting the mantra while pouring the ghee (purified butter) in the fire. The shawl of the bride and scarf of the groom were tied together and both circled the fire seven times with the groom leading, and then an additional seven times when the bride was leading. All this happened when the priest was chanting Sanskrit mantras. At the end of this, the priest declared the marriage to be a success and both families congratulated each other.

The groom was seated in the courtyard and several girlfriends of the bride and female relatives of the bride joined him to introduce themselves and make fun of him and his family and how ugly and unsightly they all looked. They were thoroughly enjoying themselves in making fun of the groom and his family. Some of them hid his shoes and would not give them back unless Rattan Lal gave to each one of them some present. The usual present was a silver ring. Rattan had been advised to get 10 or 12 silver rings and present one to each of his bride's friends and younger female relatives. This ring has a special name called *creechri*. Without these rings, the helpless groom can be in real trouble!

The ceremony, performed according to the *Sanatan Dharma* was followed by a catered lunch which was organized in a tent erected in the street in front of the bride's house. Among Rattan's professional colleagues, two of them from the Soil Salinity Research Institute of Karnal were Dr D.R. Bhumbla and Dr R.P. Abrol, and one family from IARI, New Delhi (Chaudhary Amir and Mrs Singh), also attended the lunch.

After the lunch, Rattan Lal was seated in the courtyard of his in-laws' house with his father-in-law, enjoying the warm winter sun, but also basking in the afterglow of the just-completed happy event. His father-in-law was telling him a joke about the marriage institution. He said, with a big laugh, that 'marriage is an institution to which those who are outside want to get in and those who are inside want to get out'. Rattan Lal asked him, 'But Pita Ji, you are in your eighties and are very happy to stay in the institution'. He replied, 'I got into it three times. I will tell you about it some other day.' Yes, he was married three times and was proud of his large family of 11 children. Rattan Lal's wife was the youngest and sixth child of the third wife. Whereas Rattan Lal was getting acquainted with his father-in-law, and very much admiring his English manners, friends of the bride saw the groom without a turban and without wearing his sehra or the facial cover for the first time. The bride's friends and other female members of the family were not happy to see the groom, who, to their disappointment and horror, had some grey hairs, which were now plainly visible. Rattan Lal could clearly hear these women guests telling the mother of the bride that 'the groom is an old man. We feel sorry for our friend Guddy (which was the bride's nickname and means the youngest daughter of the family; an alternative nickname also often used is Nickki, or the 'little girl'). Her friends were joking with her: 'Hai Ram mein kya kroon mujhe buddha mil gaya' (There is a famous song from the Bollywood movie called Sangam (Oh God, what should I do now that I have been married to an old man of my grandfather's age?'). While some were joking with the bride, others were more serious and complaining to the mother-in-law that Guddy has been cheated by these simple-looking but cunning villagers: 'Mother, we are very angry that you have let this happen in front of your eyes. This is not right.' The mother assured them that 'your father has done everything that is good for Guddy, and the groom has just premature grey hair. You should not tease my daughter and instead wish her well.' Furthermore, the mother also assured them that, 'this is what is in the karma of Guddy and no one could have changed the karma or her destiny. She has to live with her karma and you should bless her and accept whatever has happened.' Rattan Lal's father-in-law also heard this discussion and whispered to him in English, 'My dear Doctor Sahib (as he always called Rattan Lal in a very affectionate manner), I advise you to ignore this type of talk. They have to say something and it is a pity that they are being childish about a few of your grey hairs, and you should accept it is as a humorous welcome to the family.' From then on, Rattan Lal had very respectful and affectionate relationship with his father-in-law. They both had a lot in common. They were both educated in the western system of education, both had technical backgrounds in hydrology and water management, and both were very self-confident in their beliefs. Above all, both were self-made and came from a modest family back-ground. Thus, they both respected one another. Whenever, they were together, for about eight years after their marriage, they had a lot to talk about. Rattan Lal respected his father-in-law very much and this relationship was an important factor in the stable marriage that lasted for all of their lives.

The groom, the bride and the marriage party arrived at Rattan Lal's family home, 40 miles west of Karnal, before sunset. The bride was received ceremoniously and the sarson oil (canola oil) was poured on the doorstep to welcome the bride. In the darkness of the evening (there were no lights because of the ongoing war between Pakistan and India), three or four women, all dressed in red saris, were seated next to the bride. Rattan Lal's sister-in-law, Sheela, warned him not to call any one of them because you do not know who among them is your bride. I must warn you that they can see you very well through the veil but you cannot see that they are looking for any mistake that you make, so they can enjoy making a fool of you. Sheela advised Rattan, 'Wait for all to go away and whoever is left behind and is wearing a red dress must be your bride. Be assured that I will introduce you to her. I will not kid you and I will not pull your leg; but all of this goodness has a price.' Rattan Lal, however, was not sure if his sister-in-law could be trusted and whether she meant what she said. Nonetheless, what she said was reassuring: 'Just wait and see whatever may happen next'. In the meanwhile, Rattan Lal went back into his bedroom with a stick to be sure that there were no snakes or scorpions, and he particularly gave a thorough check to the quilt.

Sukh (that was her married name from now on) did not have a passport to enable her to undertake an international journey. Obtaining a new passport would take several weeks or months. Thus, Rattan Lal casually talked with his father about an option for Sukh to stay in the village while he went back to Nigeria by himself. His father asked him the reason and Rattan Lal mentioned that Sukh needed a passport that may take several weeks to get it issued. Rattan Lal's father made it clear that 'the place of a married woman is with her husband, and he cannot travel back to Nigeria without his wife accompanying him'. He said this loud and clear, so that not only Rattan Lal and Sukh heard it but also the entire family, especially his elder sister, brother, sister-in-law and brother-in-law. The entire family repeated the same sentence. He got the message, and thus devoted the next two weeks to procuring a passport and visa for Sukh so that she could travel with him to Nigeria. In this context, IITA management also provided a letter of support to facilitate the passport and visa processes. Because of the letter of support from IITA, these processes were completed in an orderly manner and in a short period of time.

This was another important point for the family-arranged marriage. The arranged marriage is based on the idea of union between the two families, and that is the fundamental basis of the stability of the marriage. Because the bride and groom are strangers, both families go out of their way to make them feel welcome and organize several events to familiarize them with both families. There were many dinner events and gifts from brothers, sisters, uncles and aunts from both families for the newlyweds. These events are specifically designed to increase acquaintance ('love' in western terms) between the newly wedded couple over time. In contrast, the love between the groom and bride in self-arranged marriages can be at its peak at the time of the marriage and thereafter may often be likely to decline. Thus, understanding of the love-dynamics is critical to the stability of the arranged marriage.

11.6 Arriving at IITA with his Bride

A few weeks after the wedding, Rattan Lal and Sukh left India for Nigeria via Vienna and Rome. Sukh was given an enthusiastic welcome by the IITA community. An official reception party was held in her honour at the international house, and there were also many private receptions. The reception at international house had a gift of £60 in cash, donated by his colleagues and their families, in addition to other gifts from members of the IITA community. The cash was used to get a custom-made carving at the famous curio shop in Oyo township about 30 miles north of the IITA campus. This carving is among the prized possessions of Sukh and Rattan and will remain so for life (Fig. 11.2).

Leaving her parents and relatives in India for the first time, and living in a foreign country, was difficult for Sukh. It was made particularly difficult by the fact that her husband was highly committed to his professional duties. Thus, Sukh decided to return to India in November, about one month earlier than the start of the vacation time for Rattan Lal, which was always in December. On arrival in India, Sukh told her parents that 'living in Nigeria, with a husband whose first wife is research and who is always

Fig. 11.2. The wooden carving presented by IITA faculty to Rattan Lal and Sukh on their arrival in Ibadan, Nigeria, at the end of 1971. (Photo: Lal family).

preoccupied and busy with his experiments, has not been an easy task, and I do not intend to return to Nigeria'. To dramatize this statement, she pulled the tablecloth off the dinner table and deliberately broke some dishes. However, her parents were not at all sympathetic with her decision. Her father told her, in a matter-of-fact manner, that after marriage, a wife's place is where her husband is, and she is welcome at her parents' home only for few days. She had better understand this once and for all. This message was a similar to the message Rattan received from his father about a year previous when he wanted to return to Nigeria without Sukh because she did not have a passport. Rattan's father, uneducated as he was, said that, 'After marriage, a bride lives with her husband and not with her in-laws. She must accompany him wherever and whenever he goes.' Herein lies the reason for the long duration and success of traditionally arranged marriages compared with those in the western style or the so-called 'love marriage'. The traditionally arranged marriage is, in fact, a union between the two families who have agreed to exchange their son with the daughter and vice versa. These families provide the foundation or the bond that stabilizes the marriage. That is why both families also encourage the young couple to have a child, and the sooner the better. Having a child during the first year of marriage is considered a good omen that brings happiness to the young couple and, of course, stability to the marriage through the common bond of a child.

11.7 The Lal Children and Grandchildren

Thus, when Sukh arrived in India in November 1972, both families (her parents and her in-laws) were surprised and rather disappointed that she neither had a baby nor was expecting one, and that she came back alone and was not accompanied by Rattan Lal. Not having a child after one year of marriage means that there must be some problem (physiological or sociological). Rattan Lal's bua ji also expressed her disappointment in not having a grandchild in her lap. All sisters at Sukh's family and her sister-in-law at Rattan Lal's home in the village encouraged her to have a child. Rattan Lal's family in the village arranged to give her 5 kg of ghee (purified butter) to take with her to Nigeria and to make sure that she and Rattan ate this with every lunch and dinner. They also encouraged her to cook and eat urd or black gram (*Vigna mungo*) dal soup with a spoon full of ghee in it.

Reluctantly, Sukh agreed to return to Nigeria with her husband. With a little bit of luck, and probably because of eating ghee and urd dal, Sukh was soon expecting a baby. Rather than taking a vacation in December 1973, they delayed it until May 1974 because the baby was due in March. In the meantime, Rattan Lal's bua ji was suffering from cancer and it turned out to be terminal. She was hoping to see Rattan Lal and Sukh in December. Unfortunately, she died in March during the same week that their daughter

Priya was born. Hearing of the birth, Rattan's family consoled themselves with the thought that 'We have lost one daughter and she has come back as an avatar of Rattan's daughter'.

During the next year, undoubtedly because of eating ghee and urd dal soup, Sukh gave birth to a second daughter (Pratibha or Bobby) on 11 March 1975. When Rattan Lal and Sukh came to the village in May/June with two daughters, his father said, sadly, 'I wish that these two girls were boys; that would have been God's blessing'. This was the cultural tradition – Punjabi families always preferred boys, and this tradition has affected the society for millennia. As karma would have it, the next two children of Sukh and Rattan were boys: Abhishek was born in June 1978 and Vivek in August 1979. Rattan Lal's father died of a heart attack on the same day Abhishek was born. Sadly, he did not know about the birth of his grandson. Many members of the village family remarked that Rattan Lal's father had returned as an avatar of Rattan's son.

The birth of two boys was a joyous occasion at his native village for the family and the brathery. Everyone was happy that Rattan Lal finally had a balanced family with two girls and two boys. The birth of the second son, Vivek, was often called an 'equalizer', and there was tremendous joy and thankfulness to the God for his blessing, and joyous celebrations of Lohri (spring festival). While Rattan Lal and his family were not in their native village for the celebration of Lohri, it was celebrated to honour the birth of two sons in the family. Sadly, however, there was no such celebrations of the birth of their two elder daughters. This was how the cultural traditions were during the 1970s. These have changed somewhat over the decades, but rather slowly and only slightly.

Rattan Lal and Sukhvarsha had four children (Figs 11.3 and 11.4), and they were all born in Ibadan, Nigeria, when he was employed at IITA.

11.8 Upbringing of Children and Grandchildren

The upbringing of the four children (Priya, Bobby, Abhishek and Vivek) was in strong contrast to that of Rattan Lal's. They studied at good schools, had all modern facilities available at

Fig. 11.3. The Lal family photographed in 1979 at their home on the IITA campus. From left to right: Priya, Sukhvarsha with Abhishek, Rattan with Vivek, and Pratibha. (Photo: Lal family).

Fig. 11.4. Four children of Rattan and Sukhvarsha photographed in their home at IITA in 1987 just before they left for the USA. From left to right: Pratibha, Abhishek, Vivek and Priya. (Photo: Lal family).

home, two nannies and one cook steward to look after them in Nigeria. When they were under 3, nannies would take them out on a stroller for an evening walk and there was also a children's swimming pool at the international house. This kind of household posed some challenges when the family migrated to Ohio, and had to do everything by themselves.

The family home, provided by IITA, had all the modern appliances. More importantly, it had a spacious garden, three bedrooms, one den, one living-room, a kitchen and porched screens in front of and at the back of the house. The garden had several coconut trees, bananas, grapefruit, limes and tangerines, avocados and a mango tree. But the garden also had its share of snakes, and everyone had to be careful.

The family had a cat, which belonged to the neighbour and moved in with the family after the neighbour left IITA. One day, Abhishek, when about 1 year old, was playing in the porch and the cat was also there. Rattan Lal saw that the cat was fighting with a snake and Abhishek was trying to catch it. Fortunately, the nanny saw the drama and intervened. Similar to the experience in his native village when he looked after the cattle during summer holidays, Rattan Lal also encountered snakes at IITA when he went to monitor his field experiments. In addition to green mambas, there were also spitting cobras. Thus, wearing knee-high boots and having eyeglasses were important precautionary measures.

IITA had a primary school where wives of some scientific staff were teaching primary-school children of the families living on the campus. However, the facilities and the classroom atmosphere were not adequate. Thus, IITA provided some money for tuition and two return tickets for children of IITA's scientists who would prefer to study in overseas. Similar to other students, Priya and Pratibha joined the middle-school classes at Ackworth boarding school near Leeds, UK. From the point of view of parents, especially of Sukh, nine- and ten-year-old

Fig. 11.5. Family get-together at the high-school graduation ceremony of their daughter Priya (*centre*) and Pratibha (*second from left*) in 1991. Priya is in between Vivek (12 years old) and Abhiskek (13 years old). Others in the photograph are Sukh's brother (S.P. Sharma; *back left*) and sister-in-law (Mridula Sharma; *back centre behind Priya*), their daughter (Neeru Sharma; *front right*) and son (Navneet Sharma; *front left*). (Photo: Lal family).

children staying away from home was not the best arrangement. While returning, from a summer break in Nigeria, to England in 1984, both girls had fever on arrival at Gatwick airport. They were quarantined at the airport and the parents did not know about it until a week later. Sukh urged Rattan Lal to go to England and bring both girls back. She did not want them to be at boarding school aged 9 and 10. By the time Rattan Lal arrived at the school, they had been discharged from quarantine, but were not happy with their experiences, having needles stuck in to draw blood samples several times a day.

During one of Rattan Lal's visits to Rajound for vacation from IITA, his taya (uncle) wanted to know more about his job and where he was working. Rattan Lal explained to him that he was working in Africa in a town called Ibadan. Taya asked him if this town was towards Haridwar (a Hindu religious town on the river Ganges). Rattan Lal said it was in the other direction. Then taya asked him if it was towards Mecca, and Rattan replied that it was in a different direction. Taya was surprised about the location of Ibadan, which was neither towards Haridwar nor towards Mecca. Where else could it be?

Then his taya asked him, 'Now that you have young children, what are the living conditions in Ibadan like where you live?' He specifically asked Rattan about the following: Do you have fresh milk from a cow or a buffalo? Do you have home-made ghee? Do you have C-591 wheat flour for baking the naan bread? Do you have sarson (canola) oil for cooking food? Do you have urd dal and chickpea *desi or kabuli*? Regrettably, Rattan Lal replied in the negative to each of these questions. Now taya was really concerned about Rattan Lal's negative responses regarding the availability of these basic necessities of good living as he saw it:

> 'How can you raise a young family and live in a town which does not have the most basic necessities of life? I don't understand why you need all the money that you brag about when these basic things are not available. I don't know why you accepted such a job in the first place. It is my advice that you and your young family should return to the village immediately. Now that we have a farm and a house, what more do

Fig. 11.6. Dr Pratibha Lal (*centre*) is the younger of the two daughters of Rattan and Sukhvarsha. She is in between her two daughters (*Nicki on her left and Ambika on her right*). Bela (*left*) and Juhi (*right, with cat*) are daughters of Priya Lal. This photo was taken in 2023. (Photo: Lal family).

Fig. 11.7. Jay Henry Bhakri Lal is the grandson of Rattan Lal and Sukhvarsha Lal. He is the son of Abhishek Lal and Elaina Henry. (Photo: Abhishek Lal).

> you need? Bring the children back to the house where we have milk, ghee, wheat (C-591) and sarson oil plus fresh air, open countryside and our own culture.'

This was one of the most sincere dialogues that Rattan Lal ever had with a saintly person who was down-to-earth in his thinking. His taya was really concerned that Rattan Lal's thinking was not correct and that he had been led astray by the lure of money. Rattan Lal's view, that he was working for the farmers of Africa and helping them, did not impress his taya who argued that 'such farmers also exist in their own community in Rajound and the surrounding villages, and these farmers also need help and support. You could come back and help them instead. You are being misled by the lure of money and I am disappointed with your lack of interest in working with your own community. Come back, and the sooner the better.'

While waiting for the U.S. visa, Rattan Lal was offered two jobs in India. One was as vice-chancellor (President) of an agricultural university in northeastern India. His colleagues in India advised him that he will not last long because of his lack of experience with the local politics. Earlier in the mid 1970's, Rattan Lal was also interviewed for a job at ICRISAT, in Hyderabad but at India's pay-scale. One of his British colleagues from IITA was also offered a job at ICRISAT but at an international pay-scale. The difference in the two salary scales and benefits was about ten-fold. Rattan Lal remarked to a colleague at ICRISAT that 'I can imagine someone more productive than I am by 10-20%, even 50%, but I cannot believe another researcher is 10x as productive as I am'. This system of two salary levels had been changed at all of the CGIR centers,and salaries are now comparable among researchers regardless of origin.

When Rattan Lal returned to IITA, he took two steps to do something about the situation regarding the well-being of the children and the family as a whole. One was to find a job in the USA and the other was to buy some land in Gurugram, a rapidly developing town about 12 km from New Delhi international airport with a plan to build a home there.

Bibliography

Basham, A.L. (1954) *The wonder that was India: A survey of the history and culture of the Indian subcontinent before the coming of the Muslims*. London: Sidgwick & Jackson.

12 Raising the Family in the USA

'Your children need your presence more than your presents.'
— *Rev. Jesse Jackson, Sr*

'The way you help heal the world is you start with your own family.'
— *Mother Teresa*

Abstract

Education of the children was an important issue. Both Rattan and Sukhvarsha wanted their children to be well-educated professionals. Thus, arranging their studies in the high school and university was a high priority. Therefore, Rattan Lal had to find a job in the USA and then arrange visas for the family to immigrate. This transition had some social and economic implications. However, the children stayed at their family home while studying for their college degree at The Ohio State University. Staying at home caused a disconnect with their age group and may have created difficulties in starting a successful family life of their own. Furthermore, children lost all connection with India. They did not speak any of the native languages of their parents (Hindi, Punjabi, Urdu) and were not familiar with the Indian culture. Thus, they were neither Indian nor American. This problem was faced by the first generation of families who migrated from South Asia and were settled in the western countries.

12.1 Higher Education of the Children

While Rattan and Sukhvarsha are proud of their four children (and five grandchildren by 2023), both daughters were more educated than their sons (see Chapter 11). Priya did her PhD degree from Ohio State University in human nutrition and Pratibha did her pre-medicine at OSU and completed her medical degree in surgery at Ohio Medical College in Toledo. Abhishek did his BSc degree in natural resources and MBA from the University of Maryland. Vivek obtained his BSc degree in chemical engineering. Both daughters were more accomplished professionals than the sons. Surely, Rattan Lal's father and other members of his village family would be surprised at the accomplishments of the two daughters compared with those of the two sons.

While Priya and Pratibha were completing their college degree at OSU, their mother did not want to be left behind education-wise. Thus, she also joined OSU and completed her degree in occupational therapy. Mother and both daughters were trained at OSU in the medical profession. By obtaining a BSc in occupational therapy, Sukhvarsha documented that a mother of four and wife of a professor, during her fifties, could obtain a college degree and be financially and professionally independent. All three women of Rattan Lal's immediate family did better than the three men. Women can do anything men can do, even better and more efficiently.

12.2 Finding a Job in the USA

During the early 1980s, Rattan Lal applied for a faculty job at many universities in the USA. In most cases he either did not receive any response or got a polite answer that 'competition was very tough and someone else was selected'. He had some sympathetic response from the National Soil Erosion Laboratory, at the time under the leadership of Dr Bill Moldenhauer, but it did not lead anywhere because Rattan

DOI: 10.1079/9781836990154.0012

Lal was not a U.S. citizen and thus was not qualified for a USDA job. Not being able to find a job in the USA was a humbling and also a frustrating experience. Rattan Lal was proud of his research accomplishments in Australia and especially when he was at IITA (with about 60 refereed journal articles, three bulletins, four edited books and about 30 graduate students who conducted their thesis research with him). Alas, these credentials were not enough to get an interview with any of dozens of universities and institutions to which he applied. Rattan Lal was beginning to realize the realities of finding a job in a tough and highly competitive world.

Finally, in the late summer of 1984, he received an encouraging letter from his alma mater (OSU). He was invited for an interview for two positions that he had applied for, one at Columbus and the other at Wooster. With hindsight, he knew for sure that he received the letter of invitation because members of the search committee for both positions knew him personally: two of the five members of the search committee were his former class fellows and three were his former professors and mentors. The two positions he had applied for were previously occupied by highly competent and professionally renowned professors: one at OARDC at Wooster, Ohio (as a replacement for Prof. David M. Van Doren), and the other at OSU (replacement of his major advisor, Prof. George S. Taylor). Rattan Lal was selected for an interview for both positions. These positions were advertsed for associate/full professor and Rattan Lal informed the committees that he was hoping to receive an offer for the position of full professor. Rattan Lal chose the position at Columbus, which was the replacement of his major advisor, George S. Taylor. In August 1984 he received the letter of appointment to the position within 1 year of receiving the letter. However, because he did not have major teaching experience, the position offered at Columbus was as an untenured associate professor and he was required to submit his dossier for promotion to full professorship within 2 years. This requirement implied that Rattan Lal had to submit his dossier for review within one year of his arrival at OSU. It was rather a tough situation to be in, but he had no choice and signed the letter of acceptance without knowing the tortuous and lengthy road ahead to receiving the green card for the family.

The salary and benefits for the position of an associate professor at Ohio State (US$ 48,000 p.a.) were about 40% of the total income he had at IITA (i.e. tax-free salary, additional cost of living allowance, educational support for the children, business-class ticket for the family to travel to India for a one-month vacation). The position at IITA also had a house allotted at a minimal rent and two cars were provided, one for family use and one for office use. Yet, the decision to leave this lucrative job was dictated by the educational needs of the children and the necessity that they should stay at home rather than at a boarding school. This decision was also driven by his taya's questions regarding the utter lack of basic necessities of life as per the traditional lifestyle of a Punjabi family: milk, yoghurt, ghee, dal, wheat flour, chickpeas and urd dal, among other things.

During his last year at IITA (1987), Rattan Lal was also elected President of the World Association of Soil and Water Conservation (WASWAC) and also of the International Soil and Tillage Research Organization (ISTRO). He was already Vice-President of the Commission on Continental Soil Erosion and Sediment Transport of the International Association of Hydrological Sciences (IAHS). Undertaking the new job, while also performing these extra-curricular activities, was a great challenge forhim and his young family in their new home in Columbus, Ohio.

12.3 The Green Card Dilemma

Upon receiving the letter of appointment at OSU in July 1984, Rattan Lal resigned from his job at IITA and gave a year's notice of leaving. Dr Ermond Hartman, the Director General of IITA at the time, accepted the letter of resignation. A few days after that, Rattan Lal went to the U.S. Embassy in Lagos and showed the letter of appointment from OSU, and asked for the visa application forms for him and his family. However, the staff at the embassy informed him that the wait for a visa for a passport holder from India was about 3 years and could be longer. Such a long wait period came as a big surprise to Rattan Lal who had already resigned his job at IITA.

Rattan Lal then went to see Dr Hartman about his letter of resignation, and Dr

Hartman replied that the resignation had been accepted and that it could not be revoked. Now Rattan Lal had neither a visa to the USA nor a job at IITA. With a young family, he was in a big trouble.

This was a situation from which only a divine power could rescue Rattan and his family. Dr Hartman was replaced by Dr Larry Stifel as Director General of IITA in early 1985. Dr Larry Stifel, from Cornell, with a lot of international experience, was approachable, understanding and helpful. Dr Stifel called Rattan Lal to his office, returned the letter of resignation and advised him to continue his job at IITA until the visa to the USA was approved. Yes, this was, indeed, divine intervention from the almighty Vishnu and Brahma.

By early 1987, Dr Fred Miller was appointed as the new Head of the Department of Agronomy at OSU. He reviewed the situation and realized that no one had been teaching soil physics at OSU since 1984. Therefore, en route to Makrere, Uganda, Dr Miller visited IITA in early 1987. On his return to OSU, Dr Miller wrote a letter to Rattan Lal stating that the department could wait for his arrival until the beginning of the academic year in August 1987 and no longer. They had already been waiting for about three years and the visa situation had not improved.

Rattan Lal had a phone call with Prof. George S. Taylor about the visa situation and the ultimatum from Dr Miller. Dr Taylor, very sincere and completely supportive of Rattan Lal's taking his place as Professor of Soil Physics at OSU, assured him that there was another option to expedite his visa. Dr Taylor met with Dr Francille Firebaugh, the Vice-Provost of International Affairs, and decided to approach the offices of Senator John Glenn to help with the visa process. Indeed, intervention by Senator Glenn was very effective and Rattan and the family was approved to receive the green card and report at any port of entry before 15 July 1987. The family was given 2 weeks to travel to the USA. Rattan Lal met Dr Stifel at IITA and he accepted his resignation at 2 weeks' notice, and the family were on their way to the USA. The intervention by Senator Glenn was another example of divine help of Brahma and Vishnu.

12.4 Challenges of Being a Student in the Upper Arlington School System Faced by Rattan Lal's Children

During 1987–1990, students in the Upper Arlington school system were from elite communities, and it was not easy for the Lal children to make friends. Of all the four children, Abhishek (who was nine when the family arrived in Columbus in 1987) had the most difficult time adjusting. He missed his friends at IITA in Nigeria, and was not happy in his new situation. He often returned home with bruises and signs of having had fights. Rattan Lal met with the teachers but most were unsympathetic. Abhishek was isolated, had no friends, and his lessons and homework were not satisfactory. He became rebellious and a difficult child, and continued to be so through high school and college. During the period 1987–1996, it was difficult to communicate with him.

Vivek was doing much better and was a friendly child. He was doing well at school and had made satisfactory adjustment. Vivek's behaviour was in strong contrast to that of Abhishek; he was doing well at school, was easy to communicate with, and often was the only one able to communicate effectively with Abhishek. In later life, however, the situation reversed – Abishek was easy to communicate with and Vivek was not. In strong contrast to the boys, both girls (Priya and Pratibha) were devoted to their studies.

12.5 A Big Sacrifice by the Indian Immigrant Parents

Most immigrants from India who arrived in the USA between the 1960s and 1990s have been professionally successful. These immigrants are professors, medical doctors, business leaders, politicians and entrepreneurs. They have realized the 'American dream'. Even the first-generation children of immigrants are highly educated and motivated. The Indian community stands out in terms of their hard work, dedication, excellence and, of course, success. But some of these families have made a big sacrifice in terms of the social behaviour of their highly educated and successful children and grandchildren. The fact remains that some of them have not adjusted to the new social and cultural values of the adopted home country. They are neither Indian nor American. They assume that they are American and thus do not have to follow the Indian culture and traditional family values of respecting elders and seniors. But they are also not too familiar with the American values and cultural norms. They are superfluous in American traditions and at the same time they neither accept nor respect Indian culture and traditions. In other words, this generations is socially lost and many have difficulties in getting married and having successful family lives. Considering that they are American, they neither want their parents to arrange their marriages nor are they capable of arranging marriages for themselves. While they are the big losers, the parents are disappointed and are concerned about their failure on the one hand and find themselves in a desperate and helplessness situation on the other. Those who do get married are prone to divorce, failed marriages and broken families. The third generation, or the grandchildren of the immigrants, may be better adjusted to American values and culture and integrated in the social fabric of American society. Thus, immigrant parents have made big sacrifices in trading professional success with social failure of their children.

One of Rattan Lal's sons had a girlfriend and they lived together for about 7 years, and had a child. But their relationship did not last and they were sharing the child between them. Rattan and Sukhvarsha felt very sad for the child because he did not have a family and was being shuttled between the two parents. Their second son was not able to find a wife. Sukhvarsha tried to arrange a marriage for him but he would reject all suggestions of hers. He wanted a girl of Indian heritage but with American education, culture and values. It was a lose–lose situation. Every time the son rejected the mother's advice, she would take a long breath and assure herself that she would never give up trying. For Rattan Lal, it was a very sad situation to watch.

Grandchildren of Rattan Lal's brother in his native village were another problem. They would never give up asking for money. Rattan Lal had stopped going to his native village after his father died and after his elder brother died also. Regrettably, his brother's children and grandchildren did not complete even high-school education. Thus, none of them could find a job nor did they want to practise farming. Rattan Lal felt very disappointed, frustrated and helpless but was never hopeless. He was sure that another divine intervention was due in the near future, and will happen sooner rather than later. His belief in family was strong, and that was an important factor in his maintaining a positive attitude.

13 A Faculty Member at the Ohio State University

'A good teacher can inspire hope, ignite the imagination, and instill a love of learning.'
— Brad Henry

'The Vedic traditions teaching involves the ultimate reality (brahman) and liberation (moksha) of the individual soul (atman).'
— Wisdom Library: Vedic ritual: significance and symbolism

Abstract
Rattan Lal always had a dream of being a professor at The Ohio State University, his alma mater. The dream came true in 1984 when he applied for a position vacated by his mentor who retired after about 30 years of distinguished service. After about 18 years of service at IITA, from December 1969 to July 1987, Rattan Lal arrived at OSU, on 15 July 1987. He was very enthusiastic, and felt privileged and honoured to replace his own mentor. However, the new job created some stresses which had strong implications for his health and the well-being of the family. There was the challenge of his style of teaching and the expectations that were held concerning standards. In addition, he had to learn how to deal with a different culture and different traditions, at home and at work.

13.1 Return to the Ohio State as Instructor in Soil Science

Rattan Lal and his family arrived at JFK airport, New York, at around midnight on 12 July 1987. While Sukh was sitting with four children (aged 8, 9, 12 and 13), Rattan Lal carried some 20 boxes and bags to the counter to be checked by customs and immigration. It was past midnight and the children were tired or asleep. The customs officer watched Rattan Lal bring all the bags and asked him about the passports of the family. She then asked him the reason for his entry into the USA. When Rattan Lal replied that he was joining the faculty at the Ohio State University, the officer asked him to provide her with the letter of offer. After reading the letter, she said that she would help him with his baggage. Rattan Lal was surprised, and asked her about examining the baggage, to which she replied that it was not necessary. The officer explained to him that she was herself a graduate of OSU (a buckeye) and she was sure that OSU did not appoint a crook to be a member of its faculty. On the very first day at OSU, when Rattan Lal was invited for lunch by Dr Fred Miller and Dr Frank Himes at the Faculty Club, he narrated his experiences with the immigration officer. This was the best welcome to USA that Rattan and his family could have expected. He holds fond memories of the welcome offered by the immigration officer.

Prof. George S. Taylor had arranged a five-bedroom apartment for one month at the Harrison House, which was located on Lane Avenue about half a mile from the Kottman Hall. The furnished apartment, with kitchen and basic facilities, was the most basic accommodation and was conveniently located. Prof. Taylor had also done some grocery shopping and had left some home-cooked food in the refrigerator. Professor Taylor also identified a four-bedroom house in Upper Arlington which was less than three miles from Kottman Hall. With the help of Prof. Taylor, the house and a car were purchased within one month of their arrival

DOI: 10.1079/9781836990154.0013

in the USA. Prof. Taylor made all the arrangements for the family to be settled close to OSU. Both the primary and middle schools were also near the house. The family was settled in their new home, and Prof. Taylor made sure that all needs of the young family were met. In addition to making Rattan Lal his professional successor, Prof. Taylor did everything possible to get the immigrant family settled in their new country. This indeed is an example of the great karma that Rattan Lal and his family had (Fig. 13.1).

13.2 Challenges of an Instructor

The very first class that Rattan Lal taught in soil physics, during the winter trimester from January through mid-March 1988, a total of ten weeks, was a challenging experience for him and for the students. With his reputation of his research at IITA, with a focus on sustainable agriculture through no-till farming, mulching techniques with cover cropping, and mixed cropping involving forage-based rotation and tree-based systems for integrating livestock with crops and trees, students were not expecting him to teach basic principles of soil physics involving mathematics and physics-cum-hydrology and climatology. Many students, did not understand the math-based equations and focus on quantification of different components and phases (solid, water, air and their interaction). His approach was the same in an undergraduate class on soil and water conservation and management. Frustrated with his focus on mathematics and physics in dealing with basic concepts of soil and water, some students went to the head of the department and complained about his approach to soil science. They suggested that Rattan Lal should not be allowed to use any calculus in the class and the focus should not be on mathematics. Rattan Lal was rather embarrassed when the department head told him that the students did not like his math-based approach. Rattan Lal was also surprised that most students did not have a background in mathematics, physics and chemistry, even at high-school level, which is why they had opted to study agriculture. The students' complaints

Fig. 13.1. Prof. Don Kirkham (*left*) and Prof. George S. Taylor (*second from left*) with Rattan Lal (*centre*), Dr Moses Tenyuwa and Dr Jefferey Herrick. The photograph, taken in 1989, shows four generations of soil physicists at OSU. Prof. Kirkham was the PhD advisor of Prof. Taylor, who was the advisor of Rattan Lal, who was the academic advisor of Moses and Jeffery for their PhD degrees at OSU. (Photo: Courtesy of OSU).

about both classes were regarding his math-based approach, and these were further confounded by his foreign accent. The deptartment head suggested that Rattan Lal did not teach the undergraduate class on soil and water conservation, but continue to teach soil physics, albeit without using calculus and minimizing the use of algebra and basic physics or physical chemistry.

The student evaluation of instructor (SEI) was an essential criterion for promotion and tenure of the faculty. Thus, Rattan Lal was concerned how this poor evaluation of his teaching would affect his chances of promotion to a tenured professorship, which was to happen during the next year (spring trimester 1989). Therefore, Rattan Lal focused on publications of peer-reviewed journal articles and also on writing two textbooks, one each on subjects he taught: *Principles of Soil Conservation* (Macmillan) and *Principles of Environmental Soil Physics* (CRC Press). However, writing and publication of two textbooks took several years. For example, the one on soil and water conservation was published in 1990 and that on environmental soil physics in 2004. For submitting his dossier to the Promotion and Tenure Committee, in spring 1989, Rattan Lal listed 25 journal articles published or submitted within the 18 months he had been at OSU. This was a stressful experience for him and his family. In March 1989, Rattan Lal experienced a serious heart attack and he was hospitalized for about one week. The heart attack changed his lifestyle and he had to make some drastic adjustments, which included a cardiac rehabilitation programme, begun in April 1989 and continuing through the COVID disruptions in 2020 (see Chapter 14).

Regardless of his perception as a teacher, Rattan Lal enjoyed the strong and unconditional support of his faculty colleagues and administration at department, college and university levels. Dr Fred Miller, at the college level, and Dr Firebaugh, at the university level, were very supportive. Soil science faculty (Drs Himes, Hall, Bigham, Arscott, Eckert and many others in crop sciences) were also highly supportive and positive.

13.3 Teaching Classes at OSU

Rattan Lal had a 25% teaching assignment, but interpretation of how many classes to teach varied from one three-credit class every year to one-and-a-half (one class every year and another three-credit class every other year) to two three-credit classes every year. Thus, Rattan Lal taught two three-credit classes every year from 2007 to 2023. Then there was a decision by ENR administration to assign the soil and climate class to another faculty from spring 2024.

13.3.1 Soil and climate class (ENR5268)

During 2006/07, the Vice-President of Research was Dr Robert McGrath. He launched a programme called TIE (Targeted Investments in Excellence). As a part of this programme, it was suggested that Rattan Lal, along with faculty from the School of Earth Sciences, teach a class on soil and climate. Thus, Rattan Lal and Prof. Berry Lyons started teaching the class in 2007. The enrollment was low in the beginning (5) but increased to 30 in 2023. It was a good class and both Lyons and Lal were a very good combination. However, during the fall of 2023, a decision was made by the SENR to change the class instructor. Rattan Lal was asked to find another class to teach to fulfil his 25% teaching requirements. He tried to put a team together to start teaching a 'One Health' class to begin in 2025. However, it did not materialize.

13.3.2 The one health class

There were two classes that evolved or emerged out of the research programme pursued by Rattan Lal over the 40 years of his tenure at OSU, from 1987 to 2027. These classes were Soil and Climate (ENR5268), which he taught in cooperation with Prof. Berry Lyons from the School of Earth Sciences, and the the One Health class, which he promoted around the world, but it did not materialize at OSU. The teaching of the Soil and Climate class ended when it was transferred to another faculty. However, Rattan Lal had the honour to teach this class with Prof. Lyons until Spring 2025. It was a great class. The Soil Science faculty decided, and Rattan Lal agreed, that the One Health class should be taught by someone more knowledgeable about soil fertility than was Rattan Lal. At the time of writing, the

One Health class has not been offered by anyone at OSU. Rattan Lal had hoped that it would be taught in co-operation with those interested in soil health, plant health, animal health and human health.

13.4 Co-operation in Implementation of Research and Teaching

Over 40 years of Rattan Lal's career at OSU, he was fortunate and privileged to receive numerous grants in which he brought funds to be shared with other faculty members in the SENR. Rattan Lal had been blessed and privileged to have co-operation with hundreds of colleagues from around the world, as is exemplified by many of his publications and keynote presentations. Rattan Lal presented hundreds of keynote lectures and webinars around the world, especially since 2018 (see Chapter 16). Being an OSU faculty provided Rattan Lal with a global platform to present his views on soil management, sustainable agriculture and climate change.

Rattan Lal was fortunate in receiving many awards at the department and college level. Some prominent awards received at the university level included the Distinguished Scholar Award in 1994 (Fig. 13.2), the Distinguished University Professor Award in 2010 (Fig. 13.3) and the Commencement Speaker Award in 2019, after receiving the Japan Prize (Fig. 13.4).

Rattan Lal worked on the international programme from the very beginning of his career at OSU through strong co-operation with Dr Hansen in the college and Dr Ladman and others at the Office of International Affairs. In addition, he was also President of the World Association of Soil and Water Conservation (WASWAC) from 1987 to 1989 and of the International Soil Tillage Research Organization (ISTRO) from 1988 to 1990. He organized the ISTRO conference at IITA in 1989. Thus, international activities were the central theme of his research

Fig. 13.2. President Gordon Gee presenting the Distinguished Scholar Award to Rattan Lal in 1994. (Photo: Courtesy of OSU).

Fig. 13.3. President Gordon Gee and Chair of the Board of Trustees, Les Wexner, in 2010 appointing Rattan Lal to the position of Distinguished University Professor of Soil Sciences. (Photo: The Ohio State University).

Fig. 13.4. President Michael Drake invited Rattan Lal to be the summer 2019 commencement speaker. The commencement was held at the historic St John Arena on the Ohio State University campus. This was the first time that Rattan had the honour of being a commencement speaker. (Photo: OSU).

at OSU, which was strengthened by the interest of numerous visiting scholars from around the world but especially those from India, China, Brazil, Pakistan and European organizations. He also served as a science advisor to Dr Klaus Toepfer of Germany and was strongly involved in the research programmes of his faculty colleagues in Norway, Germany, France and Iceland. From 2014 to 2020, he was President and a member of the Executive Committee of the International Union of Soil Sciences (IUSS). Thus, in 2014, the Director of SENR (Dr Bigham) had made a special provision of a university cellphone to Rattan Lal for facilitating his international work. The cellphone stayed with Rattan Lal until 2023.

13.5 The Endowed Chair

Between August 2018 and October 2020, Rattan Lal received numerous awards with a total honorarium of US$ 870,000. He added US$ 130,000 from his personal savings and donated US$ 1 million to SENR, his home department (Fig. 13.5). Following the rules of

Fig. 13.5. Rattan Lal donated honoraria of all seven international awards (US$ 1.35 million). (*left*) Rattan Lal presenting a check to President Kristina Johnson; and (*right*) to Keith DiDonato, Head of the Office of Development of the College of Food Agriculture and Environmental Sciences. (Photo: OSU).

the donation, the Dr Rattan Lal Endowed Chair was created in 2020. The title was assigned in February 2023 to a well-recognized soil scientist who was very meritorious and highly deserving of the title. The matching funds provided by the office of the President, Kristina Johnson, were also allocated to the endowed professor.

13.6 The Rattan Lal Scholar Exchange Program

Rattan Lal's goal of re-establishing co-operation between OSU, and PAU and IARI, was not met. Rattan Lal was awarded the 2024 Gulbenkian Prize for Humanity. This award of €1 million was shared among three recipients and Rattan Lal's share was €333,333. In addition, the chair of the award committee, Dr Angela Merkel, also contributed her honorarium of €25,000 among three recipients. Feeling privileged and highly blessed, Rattan Lal announced at the Gulbenkian Award ceremony in Lisbon that a new endowment will be created with this honorarium and the interest from the endowment (US$ 15,000 p.a.) would be used to sponsor a scholar exchange program among his three alma maters: Ohio State University; Punjab Agricultural University, Ludhiana; and Indian Agricultural Research Institute, New Delhi. The Rattan Lal Scholar Exchange Program was implemented from June 2025. This award was no less than a miracle and evidence of divine intervention.

Bibliography

Wisdom Library (2024) Significance of Vedic ritual. Available at: https://www.wisdomlib.org/concept/vedic-ritual (accessed 8 June 2025).

14 Heart Health

'To keep the body in good health is a duty...otherwise we shall not be able to keep our mind strong and clear.'
— Buddha

'The greatest wealth is health.'
— Virgil

Abstract

Despite being a vegetarian, teetotaler and non-smoker throughout his life, Rattan Lal has chronic heart disease and has received medical treatment for this since about 1968. This is a genetic issue that runs in the family and could have been worse if he was not following a vegetarian diet. With the professional support of three doctors (family doctor, cardiologist and electric physiologist), Rattan Lal was indeed fortunate in receiving strong medical support during his stay at The Ohio State University. Following a vegetarian diet, exercising regularly, taking medicine as advised and visiting doctors on regular basis are important principles of a successful life, even with serious heart disease.

14.1 Inherited Disease in the Family

Rattan Lal's bebe (grandmother) died of a heart attack. Rattan Lal remembered that her feet and fingers had turned blue during the last days of her life, and Rattan Lal mentioned this to his father. Unfortunately, his father also died of a sudden heart attack in 1978 on the same day that Rattan's son Abhishek was born. Rattan Lal's brother died of a sudden heart attack in 2010. His nephew (his brother's son) died of heart attack at the age of 37. Thus, heart disease ran in his family, and he was warned about it in 1959 when he went for a medical check-up to join the ACC (army cadet corps) during his first year in college in Ludhiana. Rattan Lal was also warned about his high blood pressure and palpitations during his college days at The Ohio State University, 1967–1968. When at IITA, Rattan Lal had been diagnosed with high blood pressure and was recommended to start medication in the early 1970s. The blood pressure was under control but its ramifications were evident when he joined OSU as a faculty member in 1987, and during the times when his work schedule was rather intense. He used to go to work between 6 am and 7 am when at IITA, but he started going to work at 5 am in Ohio, and his physical exercise was minimal. He was on regular blood pressure medication in Ohio but increased work intensity aggravated the risks associated with high blood pressure. Consequently, Rattan Lal had a serious heart attack in March 1989, followed by open heart surgery in 2001 and again in 2010, with an aortic valve replacement in 2022, and several stents installed during 2013–2015. He had a pacemaker fitted in 2014, and was diagnosed with congenital heart failure in 2022 and beyond. Thus, Rattan Lal had to live with chronic heart disease, which was intensified throughout his tenure at OSU. Since 2001, he has had three-monthly visits to a cardiologist and an electric physiologist. With professional support of many doctors (family doctor Dr Berry Friedman, Dr Harrison Weed, cardiologists Dr Stephen Schall and Dr John Larry, and electric physiologist Dr Raul Weiss (and later Dr Savona)), Rattan Lal has indeed been fortunate in receiving a strong medical support during his stay at OSU. In addition to diet control (low-sodium and low-fat vegetarian diet), he was regular in his daily

DOI: 10.1079/9781836990154.0014

one-hour exercise regime, which he started in 1989 after his heart attack. He joined the cardiac rehabilitation programme in 1989 and continued it until the COVID epidemic in 2020. After that, he continued his one-hour daily exercises on his own, which included either a walk or use of exercise equipment in the personal gym in the basement of his home. Good diet and exercise has become an integral part of his daily life since March 1989 and he has followed this routine religiously. In the fall of 2023, Rattan Lal had difficulty walking and doing anything strenuous (walking up a gently slope, pulling roller boards at airports, climbing stairs). During early January 2024, his cardiologist advised him to join the cadiac rehabilitation programme (4 days a week), and it was very effective in improving his strength and stamina.

This heart disease reminded Rattan Lal of the Bollywood movie *Safar*, in which the hero (Rajesh Khanna) had cancer and doctors had advised him of the seriousness of the disease. The main song of the movie is sung by the navigator of a boat: *Tum Ko Chalna Hoga. Tum na Chaloge to Chal Den gi Rahen, Manjal ko Tersen gi Teri Nigahen, Tuj Ko Chalna Hoga* ('You have to walk, and keep walking. If you don't walk, the walkway itself will start moving away from you and your eyes will long for the final destination that you cannot reach. Thus, you must keep walking'). This song became Rattan Lal's motivational mission whenever walking was a challenging task. The song and the movie, along with reciting the Gayatri mantra, were his source of energy, and they motivated him to keep walking, keep putting one step ahead of the other, come what may. The moral is to never give up, regardless of what challenges life may throw at you. It was this attitude, combined with excellent advice and care from his physicians, which was important in order to keep working and to keep engaging in international travel in his early eighties. He was proud of his work schedule – 5 am to 5 pm on workdays, and 7 am to 12 noon on weekends and holidays, which continued into his eighties.

14.2 Challenges of the Heart Surgeries

Rattan Lal was hospitalized in March 1989 and was diagnosed with a heart attack. His family, faculty colleagues and students were all surprised but supportive. Rattan Lal mentioned to his physician during his hospitalization that he did not understand the reason for his heart attack. He has been a vegetarian all his life, he has never smoked, has been a teetotaler, and has never used any substances. So why did he experience a severe heart attack at the age of 45? The doctor replied that, at this age, Mother Nature has no need for him to live. The wake-up call that Mother Nature does not need him any more made Rattan Lal think, 'What can I do that Mother Nature may find some usefulness in me and continue my life a little longer?' Thus, even a stricter control of his diet (no salt and no saturated fat) and regular one-hour exercise became more important than ever before.

Prof. George S. Taylor was concerned about Rattan Lal's heart attack. He came to see him in hospital many times. During one visit he brought with him an attorney to prepare his will. During another visit, he came when Sukh was also visiting. Prof. Taylor suggested to Sukh that she has to be prepared for Rattan Lal's heart disease. When she asked what she could do, Prof. Taylor replied that she should become financially independent by joining the university and earning a diploma that would qualify her for a professional job. He advised her to take a medical degree in occupational or physical therapy. Prof. Taylor accompanied Sukh to visit several offices at OSU and helped her to procure admission to begin her diploma in occupational therapy. The mother of four and wife of a cardiac patient completed her diploma in 1995 and worked until 2020 at a nursing home as a therapist. She retired because of the COVID disruptions in 2020. Yes, she was financially independent, and was a role model. She graduated about the same time as our two daughters, Priya and Bobby.

Prof. Taylor also introduced Rattan Lal to his family doctor, Dr Steven Schaal, who was a cardiologist. Dr Schaal accepted Rattan Lal as his patient and advised him to join the cardiac rehabilitation programme, which was supervised by him and his nurse, Joann Homan. Dr Schaal and Joann became his cardiac advisors for long time to come. He continued the programme until 2020.

Rattan Lal had open heart surgery in 2001. He recovered, and had fewer problems until 2010 when he again began to experience chest discomfort and was advised to get stents installed. However, the stents failed, probably because they were the wrong size. Thus, in 2010, Rattan Lal had to undergo his second open heart surgery. During the recovery time, Rattan Lal was experiencing palpitations and fibrillations. It was a difficult experience. This led to the installation of a pacemaker in 2014. Whereas the pacemaker, under the supervision of Dr Rahul Weiss, worked well, Rattan Lal had congestive heart failure in 2021 and it was attributed to malfunction of the aortic valve. Thus, the valve was replaced in 2021.

Two open heart surgeries followed by hospitalization, were followed by several months (36 sessions) of cardiac rehabilitation, and this became a routine that was followed on about ten different occasions between 1989 and 2020. It was followed after every surgery, angioplasty, installation of stents, valve replacement, and installation of pacemaker, and other interventions. The cardiac rehab staff and doctors were world-class professionals and always helpful and supportive. In addition, Rattan Lal always took his medication as prescribed, and never missed any, especially during travel. Hour-long exercise became the central part of his schedule while in Columbus and especially during his travels abroad and within the USA.

Following his cardiac surgery in June 2010, while he was recovering in hospital, Rattan Lal received a phone call from his colleague in Moldova saying that he was expected as a commencement speaker in about 3 weeks. The travel for this would be about 4 weeks after his surgery. Rattan Lal travelled to Moldova, stayed at the home of his host, who looked after him very well, made the commencement presentation and returned to Columbus. He did avail himself of a wheelchair in changing flights in Frankfurt and Chicago.

During his hospitalization after the second open heart surgery, Rattan Lal was angry about his experience at the hospital. The second open heart surgery was precipitated by the failure of the stents, which were supposedly installed one size smaller than what was needed. It was a technical mistake and Rattan Lal was very bitter about it. One day, an electrophysiologist, Dr Raul Weiss, came to visit him. Rattan Lal recounted that he was suffering because of someone's mistake. Dr Weiss listened and said, 'What has happened I cannot change but I can take care of you from now on if you will allow me. It is your call – be bitter and suffer or take a positive approach and let me take care of you.' Rattan Lal was grateful for his advice and Dr Weiss took good care of him. He often visited him in-between two surgeries on non-clinic days. Rattan Lal was fortunate to be attended by such dedicated and world-class professionals.

In 2014, Rattan Lal was suffering from palpitations and related complications. He also was the President-Elect of the International Union of Soil Sciences and the meeting of the IUSS was scheduled to be held in Jeju, South Korea, in 3 weeks. He went to see Dr Weiss and explained the situation to him. The decision was made that he should travel only after installation of the pacemaker. The surgery was arranged the next day so that he could travel 21 days after the surgery, from Columbus to Jeju. The surgery had to be done immediately because there were only 21 days left before travelling. When released from the hospital, Rattan Lal was instructed not to lift weights of more than 8 kg (20 lb) and not to use the left arm. His left arm and upper chest were totally blue and sore because of internal bleeding. Yet, Rattan Lal undertook the journey to Jeju on schedule, and it was among his most memorable professional trips. As the President-Elect, he made the keynote presentation to about 2500 participants. He had minor difficulty climbing the steps to the podium but no one noticed that he was getting out of breath. His former student Jian Young met him at Jeju airport and accompanied him back to the airport for his return trip to the USA.

14.3 Living with Cardiac Disease

Controlling diet, exercising regularly, taking medicine as prescribed, and visiting doctors on a regular basis are important principles for a successful life, even with serious heart disease.

During COVID, Rattan Lal installed a gym in his basement. It contained a Schwinn air dyne bicycle, a stationary bicycle, a treadmill, an upper-body exercise device (arm rotation) and a few weights. He used the gym during holidays and weekends but not after work on weekdays. In the fall of 2024, he joined the rehab programme again with exercise on 4 days every week.

14.4 Valve Replacement

During the COVID pandemic, Rattan Lal was neither able to get in-person medical advice nor follow his exercise programme. Thus, he felt week and tired and got easily out of breath. When he was finally examined, after COVID, he was found to be suffering with congenitive heart failure and both lungs were one-third full of fluid. Therefore, medication and exercise helped improve his health. However, the underlying cause of the congenitive heart failure was the clogged-up aortic valve, which had to be replaced. Once again, Rattan Lal recovered from this new cardiac problem because of the excellent facilities at the OSU hospital, superb post-surgery care during his hospital stay, world-class physicians and unconditional support from his family, especially his wife, Sukhvarsha. Rattan Lal returned to work about 1 week after his surgery.

15 The Lal Carbon Center

'We are running the most dangerous experiment in history right now, which is to see how much carbon dioxide the atmosphere can handle before there is an environmental catastrophe.'
— Elon Musk

'I think it's harder for people than it should be. But as more and more of us become carbon neutral and change the patterns in our lives to be part of the solution instead of part of the problem, we are now beginning to see the changes in policy that are needed.'
— Al Gore

Abstract

An EPA grant of US$ 30,000 p.a. started the research programme at The Ohio State University in 1988. The Carbon Management and Sequestration Center (C-MASC) was established under the auspices of the Ohio Agricultural Research and Development Center (OARDC) in 1993. In 2000, Dr Brad Moore, Vice-President, Office of Research (OR) of OSU, created a university-wide programme entitled Climate Change and Environment Research Initiative (CCERI), and appointed a four-member co-ordination committee with Rattan Lal as its chair. The CCERI support of US$ 50,000 p.a. continued for three years. In 2003, the Office of International Affairs (OIA) created the four-cluster programme across the university and one of these clusters was focused on climate change. Dr Jerry Ladman of OIA provided support of US$ 50,000 p.a. to C-MASC for co-ordinating these activities, which included the salary of an office associate to continue the seminar series initiated by OR. In 2006, Dr Robert McGrath of OR created a Targeted Investment in Excellence (TIE) programme under which C-MASC received a grant of $1.2 million for an unspecified time. Subsequent funding (2010–2027) for C-MASC was received from competitive research grants from diverse sources including: Bayer, Corteva, USDA, FFAR and other funding organizations. With the award of the 2019 Japan Prize and 2020 World Food Prize, the C-MASC was renamed by Dean Cathann Kress to CFAES Rattan Lal Center for Carbon Management and Sequestration (or Lal Carbon Center). Until 2025, the centre hosted about 420 researchers from around 60 countries who developed a co-operative training network on return to their home countries.

15.1 Research Programme on Soil Carbon Dynamics

Research on soil carbon dynamics was started at The Ohio State University with a modest grant from EPA in 1988 for a postdoctoral researcher. The initial contact with EPA was made through a workshop in Washington, D.C., to which Rattan Lal was invited to speak about the role of soil as a source or sink of carbon to the atmosphere and the effects of tropical deforestation on the terrestrial carbon budget. His presentation was based on the research done at IITA. During his last ten years at IITA (1977–1987), Rattan Lal was conducting research on loss of soil organic carbon (SOC) upon conversion of forest land to agricultural land by different methods of land development (bulldozer vs shearblade vs manual clearing, with and without burning of biomass) followed by no-till and conventional tillage methods of seedbed preparation, and with or without cover cropping with *Mucuna utilis*. At the EPA workshop, Rattan Lal explained his results on rapid depletion of SOC content

DOI: 10.1079/9781836990154.0015

upon conversion of land from forest to agro-ecosystems and the attendant degradation of soil structure leading to crusting, compaction and severe soil erosion by water. Through the contacts established at EPA, Rattan Lal received a grant of approx.. $30,000 p.a. to appoint a postdoctoral researcher and conduct some laboratory and field experiments. It was the start of a long-term programme at OSU and the beginning of a carbon management and sequestration centre.

15.2 The Carbon Management and Sequestration Center

This humble beginning led to the establishment of the Carbon Management and Sequestration Center (C-MASC) under the auspices of the Ohio Agricultural Research and Development Center (OARDC) in 1995, under the leadership of Dr Steve Slack. The C-MASC was supported by the funding from the Office of Research of the Ohio State University and by the Ohio Coal Development Office. Jackie Bird (a forester) was the Director and Bob Brown (a soil scientist) was in charge of the mine land reclamation programme of the Ohio Coal Development Office. Jackie and Bob both were graduates from OSU. Thus, both had professional interests in reclamation of mine-land soil. Subsequently, C-MASC also received funding support from the Office of Fossil Energy of the Department of Energy (DOE) based in Pittsburgh. Thus, during the formative years of C-MASC, its research programme was focused on carbon sequestration in soils of two prominent land uses and ecosystems in Ohio: (i) agro-ecosystems in relation to tillage methods, crop rotations and cover cropping; and (ii) reclaimed coal mine lands in relation to pastoral and silvicultural (forest) land uses. Field measurements of soil properties were made on age (time) chronosequences since reclamation. *In situ* field measurements were made for soil bulk density, water infiltration rate, soil moisture content, penetration resistance etc. These *in situ* measurements were complemented by laboratory assessments of aggregation, water retention characteristics, C and N content, pH, electrical conductivity, cation exchange capacity (CEC) etc. However, credible measurement of soil C stock in reclaimed mine land was confounded by contamination of soil with coal. Thus, it was important to separate fossil C from the recent plant-derived carbon. Research on coal-mined lands was also extended to Pennsylvania in co-operation with colleagues at Penn State University. Laboratory facilities for the carbon analyses (C/N; gaseous flux of CO_2, CH_4 and N_2O and identification of fossil vs new carbon through carbon dating, dissolved organic carbon) were provided by a grant from DOE.

Dr Brad Moore joined as Vice-President of Research at OSU in February 2000. He created a university-wide programme entitled Climate Change and Environment Research Initiative (CCERI), and appointed a four-member co-ordination committee (Pat Hatcher, C.K. Shum, Lonnie Thompson and Rattan Lal). He asked Rattan Lal to be the contact person for the co-ordination committee and provided a grant of $50,000 p.a. for 2 years to support the university-wide activities. Thus, a position of an office associate was created to undertake these activities which included quarterly seminars by a globally renowned scientist in climate change research. These seminars were attended by Dr Brad Moore and Dr Keith Alley from the Office of Research. Monthly meetings were also held between the co-ordination committee and Dr Moore. Thus, C-MASC had an office associate since 2000. It was also the year when Rattan Lal was appointed as a member of the Intergovernmental Panel on Climate Change (IPCC) and he was among the first-ever soil scientists to be nominated to IPCC. The Nobel Peace Prize of 2007 was shared between VP Al Gore and members of the IPCC.

Dr Brad Moore left OSU within 3 years (2003) and joined Northwestern University in Evanston, Illinois. At the same time, Dr Jerry Ladman joined as Vice-Provost of the Office of International Affairs. Dr Ladman created the four-cluster programme across the university and one of these clusters was focused on climate change. Dr Ladman provided support of $50,000 p.a. to C-MASC for co-ordinating these activities. The support provided included the salary of an office associate to continue the seminar series initiated in 2000 under the leadership of Dr Moore. Thus, there was a support from the university for 4 years of salary for an office associate and for related activities on climate change at the university level. Because of this support, C-MASC expanded its scope from soil quality and carbon dynamics to a strategy of carbon sequestration in land-based sinks for adaptation and mitigation of climate change.

In 2006, when Dr Bobby Moser was the Vice-President of Agriculture Administration and Dean of the College of Food, Agriculture and Environment Sciences, the Vice-President of Research of the Ohio State University was Dr Robert McGrath. A programme on Targeted Investment in Excellence (TIE) was created with a total funding support of $14 million. Dr Moser suggested that research on soil carbon and climate change be conducted as one of the university projects to be funded under TIE. Dr McGrath suggested that Lonnie Thompson of the Byrd Polar Center and Rattan Lal of C-MASC submit a joint proposal for funding support from TIE. The joint project was approved with a total support of $4.2 million of which $2.2 million was assigned to C-MASC. This was a great boost for the C-MASC programme and the TIE funds supported research of numerous graduate students and postdoctoral researchers for the next decade. These unconditional funds, used judiciously and entirely to support research on soil carbon dynamics, lasted for about ten years. In addition to postdoctoral researchers, research scientists and graduate students, C-MASC also attracted numerous visiting scholars from around the world. Visiting scholars came to C-MASC from India, China, Brazil, Pakistan and other countries. Some of the visiting scholars were PhD students funded through their national programme (e.g. China and Pakistan), Fulbright and Borlaug scholars (India, Algeria, Philippines), World Bank (Ethiopia) and FAO (Ethiopia). Some scholars were research scientists and faculty members whose visit to C-MASC was sponsored by their national programmes (e.g. ICAR and CSIR for India, EMBRAPA for Brazil). Visiting scholars strengthened the C-MASC programme and maintained co-operation even after returning to their home institutiona (e.g. India, China, Pakistan, Brazil). The positive and synergistic impacts of visiting scholars can never by overestimated.

Between 2000 and 2010, during the formative years of C-MASC, Rattan Lal was privileged to receive three awards which had a strong positive impact on his professional career (Table 15.1). One was the Norman Borlaug Award of TAAS presented in 2005 by Dr Borlaug and Dr Swaminathan. The second was the Nobel Peace Prize Certificate given as a member of the IPCC which shared the 2007 Peace Prize with VP Al Gore (Fig. 15.1). This award was widely recognized especially by professional organizations in Latin America and Africa. The third was the 2009 M.S. Swaminathan Award which was also organized by TAAS and presented at IARI (Fig. 15.2). These three awards were the great morale boosters and indicative of recognition of the C-MASC during the first decade of its existence.

15.2.1 The 2007 IPCC Nobel Peace Prize Award

The 2007 Nobel Peace Prize was awarded to Mr Al Gore, former Vice-President of the United States, for his book *The Inconvenient Truth*. The prize was shared with the Intergovernmental Panel on Climate Change (IPCC). Dr R.K. Pauchauri, Chair of IPCC in 2007, received the award on behalf of the committee. Members of IPCC received a Certificate of the Nobel Peace Prize (Fig. 15.1). Rattan Lal had the great honour and privilege of being invited by VP Al Gore to his farm in connection with the events on regenerative agriculture organized by him.

15.2.2 The 2009 M.S. Swaminathan Award

15.2.3 Other International Awards

The period from 2018 to 2021 was extremely auspicious for C-MASC and especially for Rattan Lal. During this time, Rattan Lal received some of the world's most prestigious awards (Table 15.1). Because of the COVID pandemic and travel restrictions, however, many of these awards, between 2020 and 2021, were received virtually.

Table 15.1. List of the international awards presented to Professor Rattan Lal between 2005 and 2024.

Year	Award	Location
2005	Norman Borlaug Award	New Delhi, India
2007	Nobel Peace Prize Certificate from IPCC	Oslo, Norway
2009	M.S. Swaminathan Award, Trust for Advancement of Agricultural Sciences	New Delhi, India
2013	Rothamsted Fellow	Rothamsted, UK
2018	Distinguished Service Medal	International Union of Soil Sciences
2018	Medal of Honor, Universidad Internacional Menéndez Pelayo	Santander, Spain
2018	GCHERA World Agricultural Prize	Nanjing, China
2018	Glinka World Soil Prize, FAO	Rome, Italy
2019	Medalist, Alumni Association	Ohio State University, USA
2019	U.S. Awasthi Award, Indian Farmers Fertiliser Cooperative Limited	New Delhi, India
2019	Japan Prize in 'Biological Production, Ecology'	Tokyo, Japan
2020	World Food Prize	Des Moines, USA
2020	Maharishi International University Award	Iowa, USA
2020	Arrell Global Food Innovation Award, Research	Ontario, Canada
2020	Ohio No-Till Council Outstanding Educator-Researcher Award	Ohio, USA
2020	Lifetime Achievement Award, Soil and Water Conservation Society, India	Dehradun, India
2021	Padma Shri Award	New Delhi, India
2021	Presidential Award, Soil Science Society of America	Madison, Wisconsin, USA
2022	Mahatma Gandhi Environment Protection Award	FAI, Columbus, Ohio, USA
2024	Gulbenkian Prize for Humanity	Lisbon, Portugal
2024	City of Columbus Ceremonial Resolution	Columbus, Ohio, USA
2025	Honorary Chair, Regenerative Agriculture	London, UK (via Solidaridad, Bhopal, India)
2025	Launch of Rattan Lal Regenerative Agriculture Award	Bhopal, India (via Solidaridad)

Rattan Lal often remarked that the reason he received most of his awards during the COVID era because much more deserving and better qualified professionals were not available. Thus, the selection committee or jury had to give these prestigious awards to someone like him because he happened to be free and able to comply with the rules and regulations of the awards.

When Rattan Lal was in Rio (Brazil) attending the 21st World Congress of Soil Science (12–17 August 2018), organized by IUSS (Rattan Lal was President of IUSS from 2017 to 2018), he received the news that he was the recipient of the 2018 GCHERA World Agricultural Prize. The World Agricultural Prize, managed by the Global Confederation of Higher Education Association for Agricultural and Life Sciences (GCHERA), is awarded through the Nanjing Agricultural University, China. Rattan also received the 2018 World Soil Prize, which is funded by Russia and presented by FAO, Rome. Receiving two world prizes one month apart was a great morale boost and extremely auspicious for C-MASC.

Fig. 15.1. (*left*) Rattan Lal was a member of the Intergovernmental Panel on Climate Change (IPCC) which shared the 2007 Nobel Peace Prize with Al Gore, former Vice-President of the United States. As a member of the IPCC, Rattan Lal was awarded the Nobel Peace Prize certificate; (*right*) Rattan Lal was invited by VP Al Gore to visit his farm near Nashville, Tennessee. VP Al Gore's farm practises regenerative agriculture based on principles of agroecology. Photo: Rattan Lal.

15.2.4 The 2019 Japan Prize

During the same period, Rattan Lal was pleasantly surprised and greatly honoured to receive news of being selected to receive the 2019 Japan Prize. Of all the awards received by Rattan Lal during the period 2018–2024 (Table 15.1), the Japan Prize was the event which was attended by the OSU administration. From the OSU administration, CFAES was represented by Dr David Benfield, Interim Director of OARDC. From the university, the Office of International Affairs was represented by Dr and Mrs Gil Latz, the Office of Research by Dr Morley Stone, and the Office of the President by Dr and Mrs Michael Drake. Rattan Lal was greatly honoured by the presence of the OSU management at the Japan Prize.

The award was presented in Tokyo in April during the week of the cherry blossom (Fig. 15.3) and in presence of His/Her Excellencies the Emperor and Empress of Japan (Fig. 15.4). One week of celebrations in Japan was the most glamorous and memorable because of the generous hospitality of the staff and meticulous planning of the sponsoring organizations of a series of events. In addition, the events also involved a reception hosted by the U.S. Ambassador to Japan (Fig. 15.5). One week of celebrations included a visit to one of the royal palaces and the associated royal gardens. The week was full of pleasant memories to last for this and many lives/avatars to come. Yes, this was the most glamorous event of our lives.

The organizers of the Japan Prize did the most commendable job and made it the most memorable event including the hotel accommodation, transport to different events, and the most generous hospitality of the hosts. Having dinner with Their Excellencies was an unforgettable and once-in-a-lifetime event. Rattan Lal was seated next to Her Excellency and Sukhvarsha with His Excellency. On the day of their departure from the hotel, the staff lined up in the lobby to say farewell.

Fig. 15.2. In 2003, Rattan Lal was presented with the M.S. Swaminathan award in Delhi, India, by M.S. Ahluwalia (*second from left*). Others in the photo (*from the right*) are Dr M.S. Swaminathan and Dr Gordon Gee, President of the Ohio State University, Dr R.S. Paroda, former Director General of ICAR (*far right*). (Photo: TAAS India)

It was an emotional departure. From Tokyo, they were escorted to Kyoto where they stayed for 1 week and were shown the cherry blossom and royal palaces, and had a traditional dinner with the Japanese geisha.

On returning from Tokyo to OSU in April 2019, President Drake invited Rattan Lal as a special guest at the spring commencement ceremony in May 2019, and as a commencement speaker for the 2020 summer commencement in St John's Arena in August 2020 (Figs 15.6 and 15.7). These were the most memorable events of Rattan Lal's professional career. From Rattan Lal's perspective, Dr Drake was a great president of OSU and highly supportive of scientific endeavours.

During the Japan Prize, the organizers also mentioned that when OSU celebrates the award, it may be pertinent to invite Mr Bill Gates. During the week of the award, Bill Gates had written an article on the importance of soil management to accomplish the Sustainable Development Goals and this article coincided with the date of the award ceremony in Tokyo. However, no such event could be organized because of the COVID pandemic and other difficulties of that time.

15.2.5 The 2019 U.S. Awasthi IFFCO Award

The 2019 U.S. Awasthi Award presented to Rattan Lal (Fig. 15.9) was the Award for Lifetime Achievement for Agriculture Development and Agriculture Research. The value of the award is Rs 2.5 million (US $30,000).

15.2.6 The 2020 Arrell Global Food Innovation Award

Awarded annually since 2018, the Arrell Global Food Innovation Award recognizes unique approaches and achievements of individuals and organizations around the globe. This award, valued at $100,000, recognizes global excellence in the broad area of food innovation and community impact. Rattan Lal received the 2020 Arrell Global Food Innovation Award. Because of the COVID pandemic, however, the event was held virtually and Rattan Lal gave the acceptance speech via Zoom. The plaque (Fig. 15.10) was received via the mail. The cheque of the

Fig. 15.3. In April 2019, during the season of cherry blossoms in Tokyo, Rattan Lal had the great honour and privilege to receive the 2019 Japan Prize in the audience of His and Her Excellencies, the Emperor and Empress of Japan. During the evening, Rattan and Sukhvarsha Lal were invited to a dinner and were seated with Their Excellencies. (Photo: The Japan Prize Foundation)

honorarium was mailed directly to the Office of Development of the Ohio State University.

15.2.7 The World Food Prize

The award of the 2020 World Food Prize to Rattan Lal was unique and highly prestigious. Whereas the Japan Prize in basic sciences is internationally recognized, not many people know about it in the USA. In contrast, the World Food Prize is the most widely recognized in the USA and elsewhere. However, the 2020 World Food Prize coincided with the COVID pandemic. Thus, Rattan Lal was humorously called the 'COVID Laureate'. Consequently, the announcement of the 2020 WFP and all related events were virtual, and the celebrations of the prestigious event were shadowed by the tragedy of COVID. Thus, the in-person award was organized in conjunction with the 2021 WFP event in honour of Dr Shakuntala Thilsted (Figs 15.11–15.13).

15.2.8 Naming of the Rattan Lal Laboratories at Punjab Agricultural University

Rattan Lal was extremely honoured by the decision of the Board of Trustees of PAU to name its agronomy and soil science building the Dr Rattan Lal Laboratories, in 2020. Because of COVID, similar to the World Food Prize ceremony, the naming ceremony at PAU was also held virtually and those in attendance had to wear face masks.

15.2.9 The CFAES Dr Rattan Lal Center for Carbon Management and Sequestration

At OSU, the WFP award celebrated with renaming the C-MASC. Dean Cathann Kress named the C-MASC as the CFAES Dr Rattan Lal Center for Carbon Management and Sequestration. The

Fig. 15.4. During the reception following the award of the Japan Prize, the Emperor and Empress had a personal audience with Rattan and Sukhvarsha. (Photo: The Japan Prize Foundation)

short or abbreviated name of the centre was chosen in 2023 as the Lal Carbon Center with the award of the C-FARM grant. Naming of the centre was a matter of great honour for the Lal family (Fig. 15.16). However, the name of the centre was also subject to the condition that it would secure funding support from external sources. This was a great motivation for Rattan Lal to stay focused and work hard to obtain funds from external sources.

15.2.10 The 2024 Gulbenkian Prize for Humanity

Similar to the great surprise about the 2019 Japan Prize, the award of the 2024 Gulbenkian Prize was also a pleasant surprise. When Rattan Lal received an e-mail around April 2024 stating that he was shortlisted for the Gulbenkian Prize and that he should provide information about his passport and other documents, he ignored it and assumed that it was a spam! He took it seriously only when he received the reminder and a video call was arranged. It was a memorable event because the award was presented by Dr Angela Merkel, former Chancellor of Germany (Fig. 15.17). The honorarium was designated for establishing the Rattan Lal Scholar Exchange Program with PAU and IARI. The first recipient of the award was the teenager activist Greta Thunberg, who gave the famous speech to the UN General Assembly ('How dare you'). Rattan Lal was the fourth recipient of this award. When someone asked if there were any similarities between him and the first recipient

Fig. 15.5. The U.S. Ambassador to Japan hosted a reception at his official residence to which the chair of the jury and other members of the organizing committee of the Japan Prize were also invited.(Photo: The Japan Prize Foundation)

Greta Thunberg, Rattan Lal responded that both of us are in the same age group! When the questioner was surprised with the answer, Rattan Lal replied that 18 and 81 are the same two numbers! It is a matter of preference whether to read from left to right or right to left! Numerical age is just a number, it does not have any other significance. The honour and privilege of having an audience with Dr Merkel gave the important impression that the world's strongest leader of her time could be so soft-spoken, caring and an attentive listener. This was indeed a 'wow' event. Meeting the President of Portugal was another memorable part of this great event.

15.3 Honorary Degrees Awarded to Rattan Lal

Rattan Lal was fortunate in receiving nine honorary degrees from prestigious universities around the world (Table 15.2). The first honorary degree was awarded by his alma mater, Punjab Agricultural University, Ludhiana in 2001. It was followed by a degree from the Norwegian University of Life Sciences at Ås, Norway, courtesy of his close friend since 1964 Dr Bal Ram Singh. The third degree was awarded from the Alecu Ruso Balti State University of the Republic of Moldova, courtesy of his colleague and friend Dr Boris Boinchin. Dr Karl-Heinz Feger, Dean of the College of Forestry and member of the Scientific Advisory Committee (SAC) of the UNU FLORES, arranged the honorary degree from the Technical University of Dresden, Germany. The award of the World Food Prize in 2020 led to the distinct honour of receiving two honorary degrees from India: Amity University, Noida, Uttar Pradesh, and from the Indian Agricultural Research Institute (IARI), New Delhi. Rattan Lal was also Honorary Professor at four universities: IARI, PAU, University of Iceland, and National University of Uzbekistan, Tashkent.

15.3.1 Ambassador of IICA on sustainable development issues for Latin America and the Caribbean region

Since 2020, Rattan Lal worked with IICA as Chair of Soil Science (LiSAM or Living Soils of America program) and Ambassador for

Fig. 15.6. President Drake invited Rattan Lal to be a commencement speaker at the 2019 summer commencement of OSU. (Photo: Courtesy of the Ohio State University)

Sustainable Development Issues. The MOU was signed by Dr Manuel Otero (along with Dr Lloyd Day and Dr Jorge Werthein) of IICA and Dr Cathann Kress (and also Dr Steve Slack and Dr Jeff Sharp) of OSU. In accordance with the co-operative programme, Rattan Lal was appointed Hon. Chair of Soil Science and Ambassador for Sustainable Development Issues of IICA. In this context, Rattan Lal presented numerous webinars to the private sector (e.g. Pepsico, Syngenta, Bayer), academic institutions (universities) and national research programmes (EMBRAPA of Brazil, ICAR of India). From 12 May 2020 to 31 December 2020, after the announcement of the WFP, Rattan Lal presented (virtually) a total of 130 webinars to many countries around the world.

This has been a mutually beneficial co-operation between IICA and the Lal Carbon Center. The topic of Living Soils America (LiSAM) is being upscaled to Africa (LiSAF) with the goal to enhance productivity in Africa while reducing the environmental effect of food systems and to soils of the tropics (LiSOT). The Green Revolution of the 1960s bypassed Africa because of degraded soils, low input of chemical fertilizers, low proportion of cropland equipped for irrigation, and lack of incentivization and motivation for farmers to adopt input-based farming systems. Rattan Lal's association with IICA was a professionally rewarding and highly fulfilling experience. He had excellent working relations with the administration of IICA. He appreciated the opportunity to work with IICA and the programme was based on mutual respect and trust.

15.3.2 World Soil Day

World Soil Day (WSD) is celebrated globally on 5 December, which is the birth date of the late King of Thailand, His Excellency King Bhumibol Adulyadej. He was very supportive of soil protection and launched a programme establishing contour hedges of vetiver grass (*Vetiveria zizanoides* or *Chrysopogon zizanioides*) to conserve soil and water. Vetiver takes its name from the Tamil term *vettiveru*, and is called *khus khus* in many parts of India. It was also on World Soil

Fig. 15.7. The 2019 summer commencement was held in St John's Arena. (Photo: Courtesy of the Ohio State University)

Day, in 2020, that the Inter-American Institute of Cooperation in Agriculture (IICA), San José, Costa Rica, signed an MOU with CFAES and initiated a co-operative programme with the Lal Carbon Center.

Rattan Lal was also invited by the Indian Soil Research Institute, Bhopal, India, to be the keynote speaker for the celebration of World Soil Day on 5 December 2024. The visit to Bhopal was hosted by Dr Sangeeta Lenka and Dr Narendra Lenka but was sponsored by the Solidaridad (South-east Asia). The Solidaridad Foundation, aimed at regenerative agriculture and soil health, has an ongoing long-term project on working with farmers and promoting regenerative agriculture. Dr Shatadru Chattopadhayay, Managing Director, Solidaridad Asia, is actively involved

Fig. 15.8. President Michael Drake invited Professor Rattan Lal to watch the Ohio State Buckeyes football game in the Presidential Suite along with dignitaries such as Mike DeWine, Governor of Ohio. (Photo: Courtesy of the Ohio State University)

Fig. 15.9. The 2019 U.S. Awasthi Indian Farmer Fertilizer Cooperative (IFFCO) India Fertilizer Association (IFA) Award was presented in New Delhi by the Founder, U.S. Awashti (*centre*). (Photo: IIFCO India)

Fig. 15.10. The 2020 Arrell Global Food Innovation Award was presented to Prof Rattan Lal by the University of Guelph in Canada. (Photo: Lal Carbon Center)

in promoting nature-based agriculture in South Asia. The Solidaridad Foundation established the Prof Rattan Lal Regenerative Agriculture Award, and appointed him to serve as Global Ambassador for Regenerative Agriculture. The 2024 World Soil Day celebrations at the Indian Institute of Soil Science in co-operation with Dr Shatadru Chattopadhayay.

On return from Bhopal , Rattan Lal was invited by TAAS and IARI to present a seminar at the Pusa campus where the Indian Society of Soil and Water Conservation presented to him the 2020 Life Time Achievement Award (Fig. 15.18).

15.3.3 The 2018 Medal of Honor from UMIP

Rattan Lal had a very good co-operative relationship with Dr Ildefonoso Pla Sentis since the time when the latter worked at the University of Maracay, Venezuela. Along with Dr Eric Roose of ORSTOM, Montpellier, three of them attended the meetings of the International Soil Tillage Research Organization (ISTRO) in which Dr Pla was Vice-President for a long period and he had done exemplary work on soil erodibility in relation to soil physical properties using field-operated rainfall simulator (Fig. 15.19).

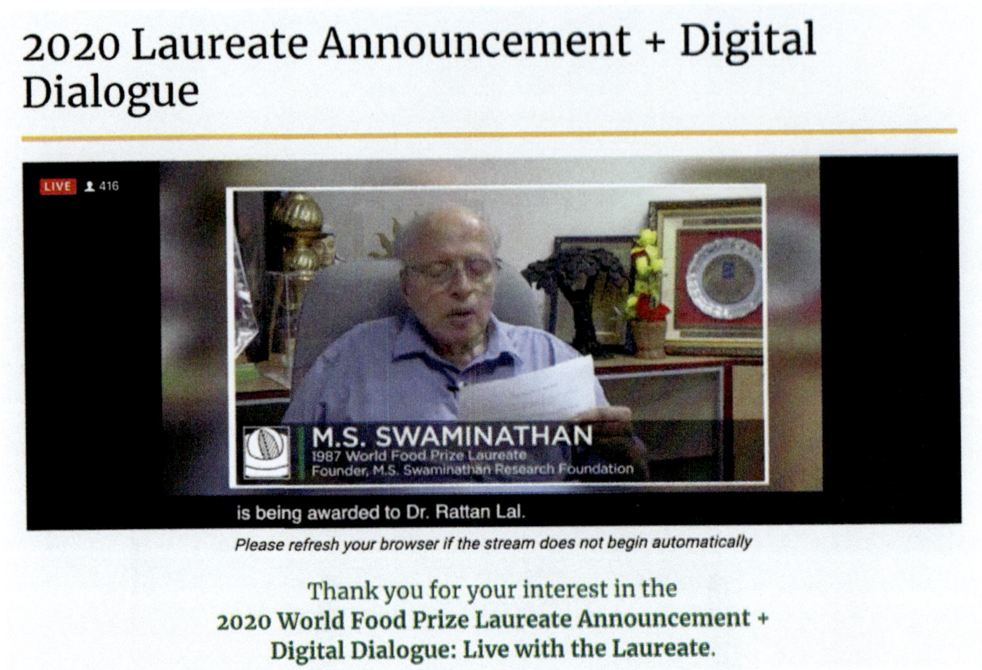

Fig. 15.11. Because of COVID, the 2020 World Food Prize event was virtual. The announcement of the award to Prof Rattan Lal was made by the Chair of the Award Committee, Dr M.S. Swaminathan. (Photo: World Food Prize, Des Moines, Iowa)

15.4 Recognition Around the World

The research programme of the Lal Carbon Center was recognized around the world but especially in Latin America, probably because of the close association with IICA. It is also important to state that 99% of accomplishments, or lack of them, depend on the focus, dedication, determination and willpower of the individual along with the support from the department, college and university. With the support and good will of the home institution (SENR, CFAES and OSU), and with determination and willpower, one is in control of one's own destiny.

15.5 The Overall Experience at the Ohio State University

Overall, Rattan Lal had a great experience at OSU during the 40 years of his tenure from 1987 to 2027. He always moved forward and achieved his goals and mission because he stayed focused, committed, determined and on-track for the task ahead. His mission always has been to advance and promote the science, practice and implementation of sustainable management of soil, and of transforming agriculture into a part of the solution to address environmental issues of the 21st century.

Rattan Lal also received respect and recognition in Europe, Asia, Latin America and the Caribbean region. It is important to note that several administrators (see Acknowledgements) were highly supportive and helpful. There was no interference in his research and outreach programme by any administrator at department, college or university level.

Therefore, Rattan Lal's scientific programme always moved ahead at full speed and the positive attitude of all was encouraging. Rattan Lal always drew strength from his congenial work environment. This was the basis of his oft-cited remark: 'I have received all these awards and recognition because I am affiliated with the OSU.

Fig. 15.12. The 2020 World Food Prize was presented to Rattan Lal in 2021 in conjunction with the 2021 World Food Prize event held in person following the ease of COVID travel restrictions. The award was presented by the grand-daughters of Mr Ruan and Dr Norman Borlaug. (Photo: World Food Prize, Des Moines, Iowa)

Fig. 15.13. During the World Food Prize ceremony in 2021, when Dr Shakuntala Haraksingh Thilsted received her award, a surprise announcement was made by the Secretary of Agriculture, Hon. Tom Vilsack (Fig. 15.14) to recognize Rattan Lal's receiving of the 2020 World Food Prize because there was no in-person presentation in 2020. (Photo: World Food Prize, Des Moines, Iowa)

I am whatever I am because of the OSU. My professional identity is based on affiliation with the OSU, and am not effective without it.' This positive and optimistic attitude was the basis of his professional integrity and moral strength.

15.6 The Role of Support Staff, Graduate Students and Visiting Scholars from Around the World

Another important reason for Rattan Lal's success was his attitude towards his staff, students, researchers and visiting scholars. He was always proud of them, and believed that his staff, students and research scholars were the best among the best. He had an opportunity to work with more than 400 scholars from around the world. They contributed tremendously to the scientific programme, and made the Lal Carbon Center a globally recognized institution. Because these scholars came from different parts of the world, representing 60 countries from six continents, the Lal Carbon Center was much bigger than it would seem (see maps in Figs 15.20–15.24).

The international aspect of Rattan Lal's programme is documented in the countries of origin of his graduate students (Fig. 15.20), postdoctoral scholars (Fig. 15.21), visiting scientists (Fig. 15.22) and countries in which he presented webinars (Figs 15.23 and 15.24). These countries are also outlined in Tables 15.3 and 15.4.

One of the visitors to the centre looked at the majestic Kottman Hall and said, 'so, this building is the Lal Carbon Center'. Rattan Lal replied that Kottman Hall is only one of the buildings; there are many more throughout all continents and many countries. Wherever, former students and research scholars (400 of them) were based, that building was an integral part of the Lal Carbon Center. Yes, the sun never sets on the co-operative workplaces of the Lal Carbon Center.

Rattan Lal was privileged to have staff, students, postdocs, research scientists and

Fig. 15.14. The U.S. Secretary of Agriculture, Tom Vilsack, congratulating Rattan Lal following his surprise announcement about the 2020 World Food Prize award which was not presented in person because of the COVID pandemic. Sukhvarsha Lal (wearing the mask) is in the background. (Photo: World Food Prize, Des Moines, Iowa)

visiting scholars who proved to be the best among the best in their profession. To Rattan, there neither was, nor will be, as good and as dedicated staff as they. These, totalling around 430 from around the world, were the major strength of the centre (Tables 15.3 and 15.4). For Rattan Lal, his staff were the avatars of gods and goddesses. Indeed, he often remarked that he would be nothing without them. It is difficult to mention all names but three of those who got him started in 1987 were Ken Scaife, lab/field manager; Karla Gutheil, office associate, and Jeff Herrick, the first PhD student. These three were dedicated professionals, and got Rattan Lal started on a great path of progress and success. Some other helpful field and lab staff are listed in the Acknowledgements. Indeed, Rattan Lal was privileged to have staff, students, postdocs, research scientists and visiting scholars who proved to be the best among their profession.

15.6.1 Global webinars and presentations

Over his career, Rattan Lal travelled to 103 countries, and some of these (e.g. Brazil, Costa Rica, Germany) three or more times per year. In addition to the in-person visits, seminars were also presented virtually, especially during and since COVID. The total number of keynote presentations over the 50 years from 1974 to 2025 is about 1000, and most of these are depicted in Figs 15.23 and 15.24 and outlined in Table 15.5.

15.7 Co-operation with USDA-NRCS and ARS

With strong support and encouragement from Horace Smith, Chief of the Land Use and Soil Survey Division of NRCS, Rattan Lal was honoured and privileged to have a long-term co-operative programme with Dr John Kimble (NRCS) and Dr Ron Follett (ARS). This co-operation between Ohio State, USDA-NRCS and USDA-ARS was funded by NRCS courtesy of Horace Smith. This team (Lal, Kimble and Follett, Fig. 15.25) together organized about 18 conferences between 1995 and 2010 (about one each year) on carbon sequestration in soils of the U.S. cropland, grazing land, forest land, mine land, tundra, arid regions and humid tropics, and also on policy issues. These conferences were managed by Lynn Everett who was very

Fig. 15.15. The soil science agronomy building at PAU was named the Dr Rattan Lal Laboratories in 2020 during the COVID pandemic. (Photo: PAU)

Fig. 15.16. The C-MASC was renamed by Dean Dr Cathann Kress to CFAES Rattan Lal Center for Carbon Management and Sequestration (Lal Carbon Center) following the 2020 World Food Prize award. The Punjab Agricultural University also named the soil and agronomy building as the Dr Rattan Lal Laboratories (Fig. 15.15). (Photo: Lal Carbon Center)

well organized. This long-term co-operation created and strengthened the foundation of the Carbon Management and Sequestration Center (later the CFAES Rattan Lal Center for Carbon Management and Sequestration) and the Inter-American Institute of Cooperation in Agriculture.

15.8 Co-operation with International Scientists and Leaders in Activities

Being affiliated with IITA and OSU created strong networks that benefitted Rattan Lal in developing a global programme on soil management and carbon sequestration.

Fig. 15.17. The 2024 Gulbenkian Prize for Humanity was presented by Dr Angela Merkel (*right*), the former Chancellor of Germany, and the President of the Board António M. Feijó (*left*) on 11 July 2024 in Lisbon, Portugal. The honorarium from the award was donated to Ohio State to create an endowment to sponsor a scholar exchange programme with the Punjab Agriculture University, Ludhiana, and the Indian Agriculture Research Institute (IARI). (Photo: Gulbenkian Prize Foundation)

Rattan Lal was highly blessed, privileged and fortunate in receiving some highly prestigious international awards (see Table 15.1). Several colleagues from the Ohio State University (School of Environment and Natural Resources; Office of International Program in Agriculture of the College of Food, Agriculture and Environment Sciences; Office of Research and Office of International Affairs, and others) prepared the nomination and submitted these. Several colleagues from the USA and internationally provided letters of support. Institutions in India and elsewhere were very thoughtful and supportive in nominating him for several awards.

15.8.1 The 2013 Rothamsted Fellow

The Rothamsted Experimental Station was founded in 1843 by John Bennet Lawes, a noted Victorian entrepreneur and scientist who had founded one of the first artificial fertilizer manufacturing factories in 1842, on his 16th-century estate, Rothamsted Manor, to investigate the impact of inorganic and organic fertilizers on crop yield. Rattan Lal had the good fortune to visit the station several times, but especially in 2013, when he was made Rothamsted Fellow. Fig. 15.26 shows Dr David Powlson with Dr David Jennings, both of whom worked on these long-term plots.

Table 15.2. Honorary degrees awarded to Prof Rattan Lal.

Year	Degree and institution
2001	Doctor of Science, Honoris Causa; Punjab Agricultural University, Ludhiana, India
2005	Doctor of Science, Honoris Causa; Norwegian University of Life Sciences, Ås, Norway
2010	Doctor of Science, Honoris Causa; Alecu Russo Balti State University, Republic of Moldova
2015	Doctor of Science, Honoris Causa; Technical University of Dresden, Germany
2017	Doctor of Science, Honoris Causa; University of Lleida, Spain
2018	Medal of Honor, UIMP, Santander, Spain
2018	Doctor of Science, Honoris Causa; Gustavus Adolphus College, St. Peter, Minnesota
2019	Doctor of Science, Honoris Causa; Pontifical Catholic University of Valparaíso, Valparaíso, Chile
2020	Doctor of Science, Honoris Causa; Amity University in Noida, Uttar Pradesh, India
2021	Doctor of Science, Honoris Causa; Indian Agricultural Research Institute (IARI), New Delhi

15.8.2 The 2018 GCHERA World Agriculture Prize

Rattan Lal was the President of IUSS in 2018. In addition to presiding over its congress in Rio, he also received an honorary degree from Gustavus College, Minnesota, and the Medal of Honor of UIMP in Spain. The news about the 2018 GCHERA World Agricultural Prize held in Nanjing, China (Fig. 15.27) was received when he was in Rio.

15.8.3 The 2018 Glinka World Soil Prize

The Glinka World Soil Prize is an annual prize awarded since 2016 to researchers for their direct contributions to preserving the environment, food security and poverty alleviation. The winner receives a US$ 15,000 cheque and a Glinka gold-plated medal.

The prize is named in honour of the Russian soil scientist Konstantin Glinka (1867–1927). It is awarded by the Global Soil Partnership (GSP) of the United Nations Food and Agriculture Organization (FAO), Rome. The award is generally presented to corporations and Rattan Lal was the first scientist to receive the award (Fig. 15.28).

15.8.4 The 2018 Medal of Honor from UMIP

Rattan Lal had a very good co-operative relationship with Dr Ildefonoso Pla Sentis since the time when the latter worked at the University of Maracay in Venezuela. Along with Dr Eric Roose of ORSTOM, Montpellier, the three of them attended the meetings of the International Soil Tillage Research Organization (ISTRO) in which Dr Pla was Vice-President for a long period and had done exemplary work on soil erodibility in relation to soil physical properties using a field-operated rainfall simulator. Dr Pla Sentis invited Rattan Lal to receive an honorary degree from the University of Lleida, Spain. The year 2018 was indeed a very auspicious year for Rattan Lal. In addition to several other awards, he was also presented the Medal of Honor by Dr Pedro Duque, Rector of Universidad Internacional Menéndez Pelayo (UMIP) in Santador, Spain (Fig. 15.29).

15.8.5 The 2022 Ghandi Environmental Award

Rattan Lal has received many awards by associations of Indian communities. Whenever Rattan

Fig. 15.18. The 2020 Lifetime Achievement Award of the Indian Society of Soil and Water Conservation was postponed due to restrictions on meeting during COVID-19 and was presented to Rattan Lal on 11 December 2024 during the seminar event of the Alumni Association of IARI held at Pusa Institute, New Delhi. (Photo: TAAS India)

Lal received an international award, he was also honored at an event organized by the Indian community of Central Ohio. The community also invited many delegations from India (e.g. PAU, Minister of Agriculture, etc.), and Rattan was invited to these events.

In 2022, the Gandhi Memorial Society had organized a centennial celebration of the birth of Mahatma Gandhi. The guest of honour was R.M. Gandhi, grandson of Mahatma. Rattan Lal received from him the 2022 Mahatma Gandhi Award for Environmental Protection.

15.9 The Lal Carbon Center at the National University of Tashkent

In March 2025, Rattan was pleasantly surprised to receive a letter from the Rector of the of the National University of Uzbekistan, Prof Dr I. Madjidov, stating that on the 2025 World Environment Day, the National University of Uzbekistan will inaugurate the Carbon Center named after Prof Rattan Lal. Rattan Lal attended the event in person in Tashkent on 22 April 2025 (see Figure 15.31).

15.10 Retirement

Rattan continued working through his early eighties following a schedule of 5 am to 5 pm on workdays and 7 am to 12 noon on weekends and holidays. He did not take any sick leave nor vacations. To him, going to work at 5 am was a therapeutic experience. Being at his desk at 5 am was a matter of pleasure and an important part of his being. He used his early time in the office for planning his research and schedule. This work schedule also kept him focused, especially on the international issues of research, outreach and teaching.

Rattan Lal did not believe in retirement; he wanted to continue working and travelling internationally as long as he could. He believed that the best way to maintain functionality of

Fig. 15.19. (*top*) Dr Ildefonso Pla Sentis, Dr Eric Roose and Dr Rattan Lal attended the ISTRO conferences from the very first one held at IITA in 1975; (bottom) Dr Pla Sentis invited Rattan Lal to receive an honorary degree from the University of Lleida, Spain. The year 2018 was indeed a very auspicious year for Rattan Lal. (Photo: Dr Pla Sentis)

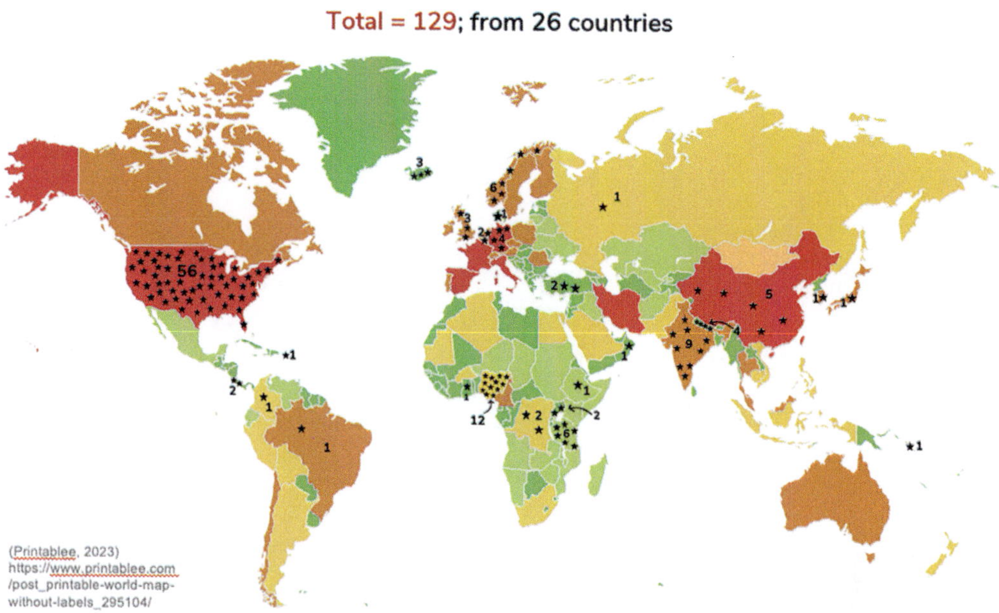

Fig. 15.20. Graduate students at the Lal Carbon Center from around the world.

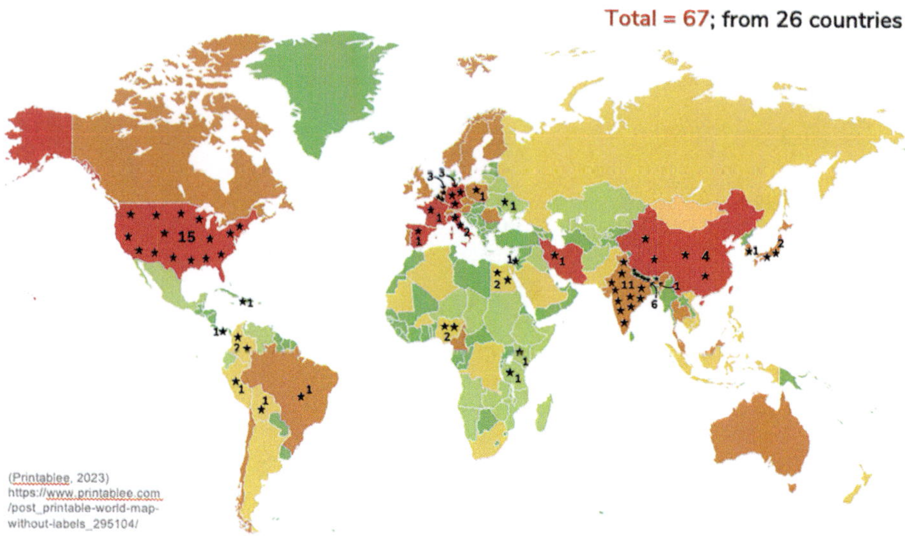

Fig. 15.21. Postdoctoral fellows and research associates at the Lal Carbon Center from around the world.

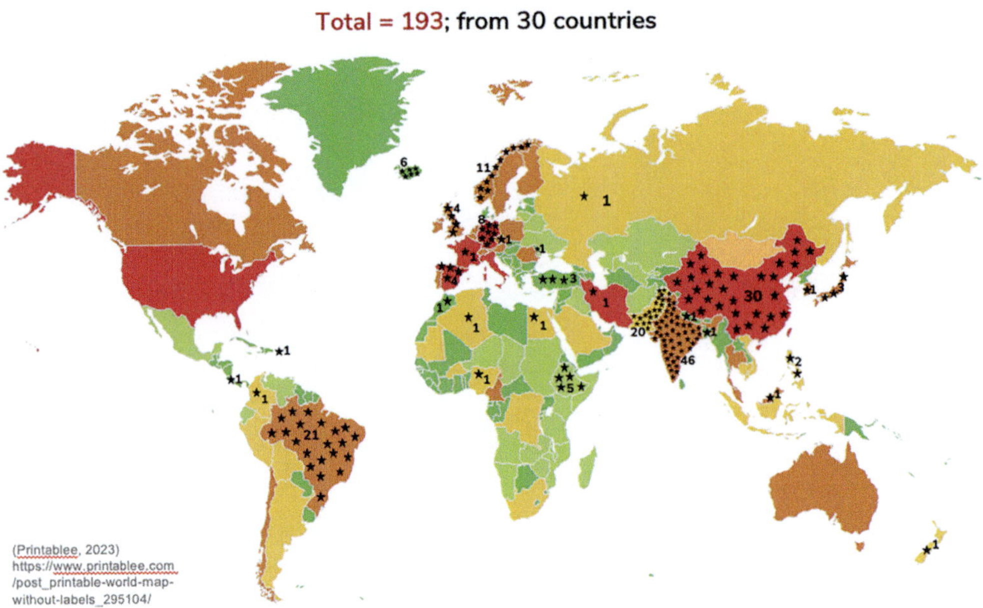

Fig. 15.22. Visiting scholars at the Lal Carbon Center from around the world.

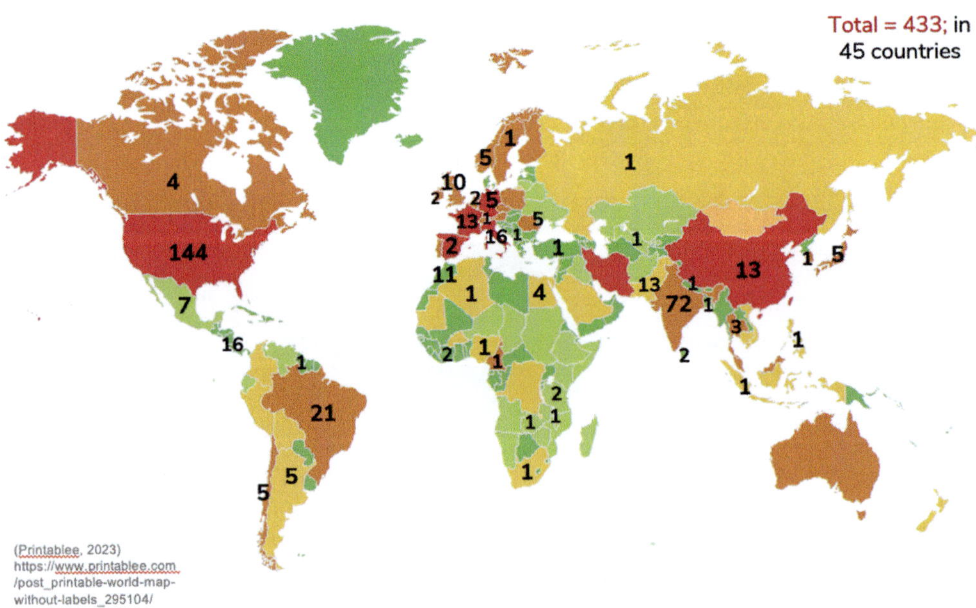

Fig. 15.23. Presentations given around the world from 1974 to 2017.

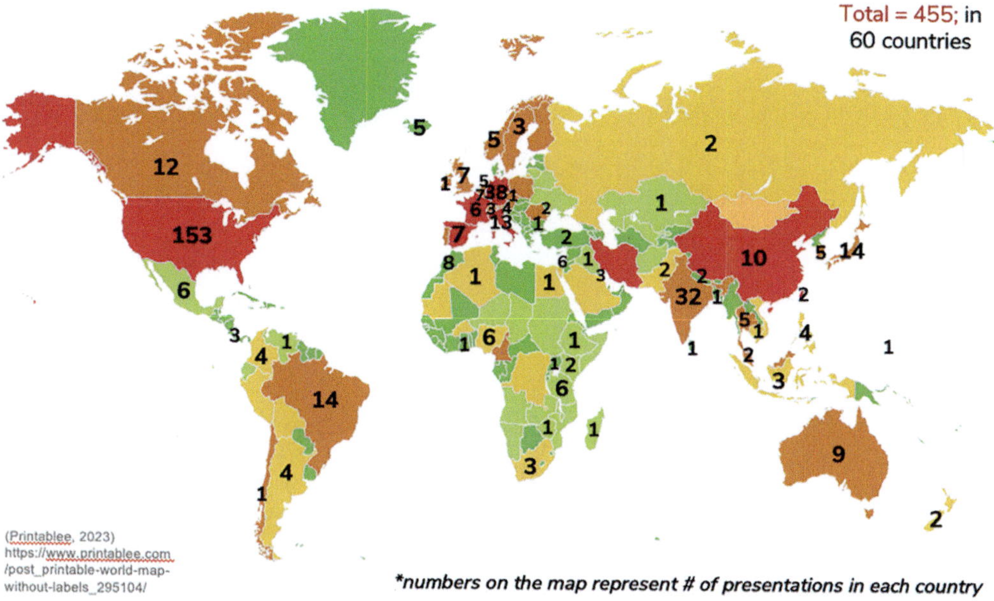

Fig. 15.24. Presentations given around the world from 2018 to 2024.

Table 15.3. Decadal distribution of mentees received at the Lal Carbon Center.

Scholar	1970s	1980s	1990s	2000s	2010s	2020s	Totals
Graduate & Honours Undergraduate Students	8	24	22	23	45	32	154
Post-doc and Research Associates	3	5	10	23	14	13	68
Research Scientists	0	0	0	5	4	6	15
Visiting Scholars	0	2	15	48	112	16	193
Total	11	31	47	99	175	67	430

Table 15.4. Global distribution of scholars received at the Lal Carbon Center.

Scholars	Africa	China	India	Brazil	Pakistan	Europe	Nepal	Bhutan	Other	Totals
Graduate and Undergraduate Honours Students	25	4	8	1	0	19	5	0	92	154
Post-doc and Research Associates	5	4	11	1	0	12	6	1	28	68
Research Scientists	2	0	3	1	0	1	3	1	4	15
Visiting Scholars	9	30	65	20	20	36	1	0	12	193
Total	41	38	87	23	20	68	15	2	136	430

Table 15.5. Countries of keynote presentations.

Country	Number	Country	Number
USA	307	Mexico	16
India	108	Spain	9
Germany	43	Thailand	8
Italy	30	Tanzania	8
Brazil	37	Argentina	9
China	24	Nigeria	7
Costa Rica	20	The Netherlands	7
France	19	Belgium	7
Japan	19	South Korea	6
UK	18	Chile	6
Canada	16	Iceland	5
Pakistan	16	Philippines	5
Morocco	25	e.g. ypt	5
Norway	10	Indonesia	4
Australia	9	South Africa	4
Other	118		
Total	**925**		

Fig. 15.25. Rattan Lal worked with Dr John Kimble of NRCS and Dr Ron Follett of ARS from 1993 to 2010. Together, they organized 18 international conferences and edited books on soil carbon sequestration as proceedings of these conferences. Output of these conferences formed the basic foundation of the Carbon Management and Sequestration Center (C-MASC), later named as the CFAES Rattan Lal center for Carbon Management and Sequestration. (Photo: Lal Carbon Center)

Fig. 15.26. In 2012, Rattan Lal was elected Rothamsted Fellow and visited the historic long-term agronomic experiment started in 1850. Dr David Powlson (*right*) and Dr David Jennings (*centre*) both worked on these long-term plots. (Photo: Rattan Lal; long-term experiment at Rothamsted, UK)

hands, feet and brain was to continue working as long as health would permit. In a private and informal conversation, a colleague asked Rattan Lal when he planned to retire. In response, he said 'I do not know'. To Rattan Lal's response, his colleague asked 'Do you plan to keep studying soil until you become soil yourself?'. Although Rattan Lal did not think much about this question previously, he thought that those who could accomplish this life's mission were indeed lucky and privileged. Availability of excellent health support at the Ohio State was an important factor in his continuation of an active work schedule and research programme. If he had a choice, Rattan Lal would encourage everyone to stay active as long as they can. Retiring early could be useful if there was another job to do, but staying home and sleeping most of the time was no good for him.

Rattan Lal always gave high priority to exercising 1 hour each day for 5 days or more per week. When the season permitted (spring, summer and fall), he preferred walking at the campus or in the neighbourhood where he lived. In addition, he had workout facilities in his basement, which included air dyne, stationary bike, treadmill, upper-body exercise equipment and weights.

Fig. 15.27. (*top*) In 2018, the GCHERA World Agriculture Prize was presented to Rattan Lal by John Kennelly, former President of GCHERA, at a ceremony on the campus of Nanjing Agricultural University, China; (*bottom*) a photo taken after receiving the GCHERA World Agriculture Prize in 2018. (Photo: GCHERA)

Fig. 15.28. (*top*) During the 2018 Glinka Soil Prize, Rattan Lal made a presentation on the importance of soil science to advance the Sustainable Development Goals (SDGs) of the UN and the role of the FAO and other UN organizations in advancing the cause of soil science. (*bottom*) The 2018 Glinka World Soil Prize was presented to Rattan Lal by Maria Helena Semedo, Deputy Director General, Climate and Natural Resources, FAO, Rome. The award is funded by the government of Russia. (Photo: Courtesy of FAO, Rome, Italy)

Fig. 15.29. In 2018, Rattan Lal was presented with the Medal of Honor by Dr Pedro Duque, Rector of Universidad Internacional Menéndez Pelayo (UMIP), in Santador, Spain. (Photo: UMIP)

Fig. 15.30. The 2022 Mahatma Gandhi Award for Environmental Protection was presented by the grandson of Mahatma Gandhi, R.M. Gandhi, at an event organized by the Gandhi Memorial Society of Central Ohio, Columbus, Ohio. (Photo: Gandhi Memorial Society)

Fig. 15.31. *(top)* The ribbon cutting ceremony to inaugurate the Lal Soil Health and Carbon Management Center at the National University of Uzbekistan, Tashkent on 12 April 2025. *(bottom)* Rattan Lal and others looking at the modern equipment within the Lal Soil Health and Carbon Management Center at the National University of Uzbekistan, Tashkent. (Photo: Prof L Gafurova, Soil Science Department)

Bibliography

Brinkley, D. (2013) Joe Biden: The *Rolling Stone* interview. Available at: https://www.rollingstone.com/politics/politics-news/joe-biden-the-rolling-stone-interview-204720/ (accessed 10 June 2025).

16 Global Research, Educational and Outreach Programmes of the Lal Carbon Center

'Climate *change knows no borders.'*
— Angela Merkel, in a speech at the Lowy Institute in Sydney, Australia

'...the world is reaching the tipping point... we risk denying present and future generations the right to a healthy and sustainable planet – the whole of humanity stands to lose.'
— *Kofi Annan, interview with* The Guardian

Abstract

Rattan Lal developed an interest in international activities in 1963 when he started his MSc degree programme at IARI, received graduate assistantship from the Rockefeller Foundation, and his academic advisory committee consisted of two soil scientists, one each from the Rockefeller Foundation and the US Agency for International Development (USAID). Earning a PhD from the Ohio State University in 1968, followed by a postdoctoral programme at the University of Sydney (1968–1969), and his position of soil physicist at the International Institute of Tropical Agriculture (IITA), Ibadan, Nigeria (1969–1987), accentuated his interests in sustainable management of soil and agriculture at a global level with focus on soils of the tropics. His interest in international programme was strengthened when he joined the Ohio State University, which had a strong record of international research and education at department, college and university level. The Dean of CFAES and Vice-Provost of Office of International Affairs also promoted the international programme with moral and logistical support. Being a faculty member at The Ohio State University was an important factor that supported his interest in international programme. By 2025, Rattan Lal had visited 103 countries.

16.1 Interests in International Activities

Rattan Lal's international career started in 1963 with his appointment as a research assistant, at Rs 200 per month, from the Rockefeller Foundation, for pursuing his MSc degree in soil science at the Indian Agricultural Research Institute (IARI) in New Delhi. Of the five students admitted into the MSc programme in soil science, Rattan Lal was the only one who received the assistantship from the Rockefeller Foundation. The Director of the Rockefeller Foundation in New Delhi at the time was Dr Ralph Cummings Sr, and his team comprised Dr Bill C. Write and Dr Guy Baird as soil scientists and agronomists. Thus, both Dr Write and Dr Baird served on the academic advisory committee of Rattan Lal for his MSc degree. These contacts were important to Rattan Lal's future career including his admission to the Ohio State University for his PhD (1965–1968); appointment as Senior Research Fellow at the University of Sydney (1968–1969); and appointment as a soil physicist at the International Institute of Tropical Agriculture (IITA) (1969–1987). While working for IITA, the geographical scope of Rattan Lal's research comprised the humid and sub-humid tropics. Thus, he had a unique opportunity to travel and conduct research in the humid and sub-humid tropics of Africa (Nigeria, Ghana, Ivory Coast, Sierra Leone, Cameroun, Congo, Zaire), Asia (Malaysia, Indonesia, Thailand); and South

DOI: 10.1079/9781836990154.0016

America (Brazil, Columbia, Peru), and elsewhere throughout the world.

These international activities were continued upon his appointment at OSU as Professor of International Soil Sciences, and later as Director of the Carbon Management and Sequestration Center, which was renamed in 2020 the Lal Carbon Center.

16.2 Graduate Students and Visiting Scholars from Around the World

The international programme and activities were strengthened with strong support from Dr Bobby Moser, the long-time Dean of the College of Food, Agriculture and Environment Sciences, and Dr David Hansen, Associate Dean and Director of the International Program in Agriculture (IPA) at OSU. Because of the strong encouragement of the CFAES and IPA, Rattan Lal mentored many graduate students from around the world who conducted thesis research in their country of origin. Additionally, students from the USA also conducted thesis research abroad. The international focus also attracted many visiting scholars (India, China, Brazil, Pakistan) who completed a part of their degree programme in the laboratory of the Lal Carbon Center at OSU.

A long-term programme was developed with the Norwegian University of Life Sciences from the mid-1990s. The co-ordinator of the programme from NMBU was Prof. Bal Ram Singh (Fig. 16.1), who was also a visiting scholar at OSU.

Fig. 16.1. Prof. Bal Ram Singh of the Norwegian University of Life Sciences (*right*) developed a long-term co-operative programme with the Lal Carbon Center. This picture was taken in 2018 in Rio, Brazil, with Prof. Flavio A.O. Camargo, Chair of the organizing committee of Brazil for the Congress of the International Union of Soil Sciences, while Prof. Rattan Lal was the President of the IUSS. (Photo: Prof. Bal Ram Singh).

Table 16.1. Authorship of journal articles as of August 2025.

Authorship	1960s	1970s	1980s	1990s	2000s	2010s	2020-2024	Totals
First	2	22	47	52	89	62	41	315
Second	0	13	66	40	151	148	22	440
Third	0	2	2	13	17	111	29	174
Others	0	1	0	10	10	105	90	216
Totals	2	38	115	115	267	426	182	1145

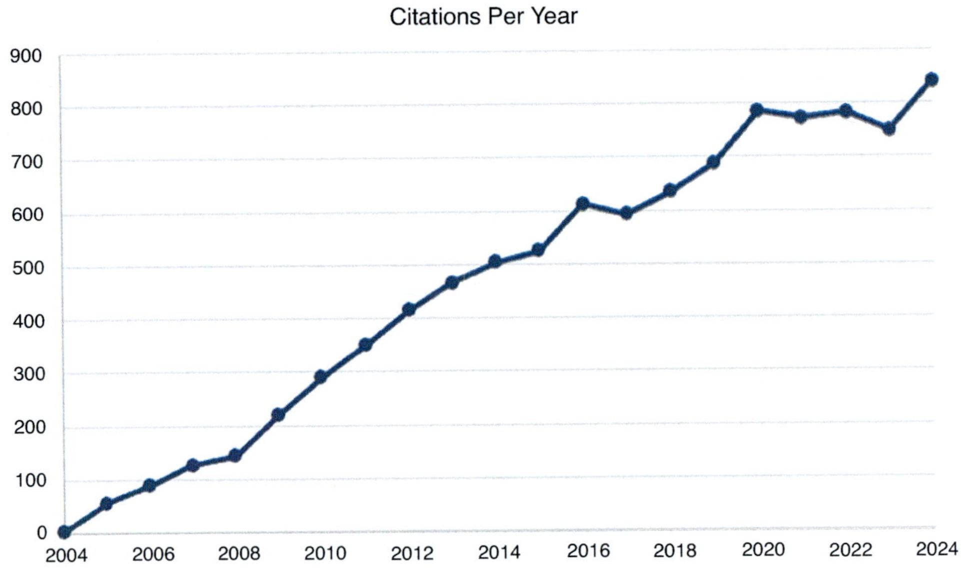

Fig. 16.2. 'Soil carbon sequestration impacts on global climate change and food security' (*Science* 304, 1623–1627) citations (2004 is year 0).

Rattan Lal and Bal Ram Singh were graduate students at IARI during 1964–1965. With funding support from NORAD, Dr Singh had numerous projects in Africa and he invited Rattan Lal to be a co-operator and co-supervisor of many students from Africa. It was a great collaboration.

16.3 Total Citations

With 430 co-operators from around the world, Rattan Lal had written more than 1100 refereed journal articles. Thus, he had total research citations of 182,000 by the end of August 2025. One of his articles published in *Science* in 2004 reached citations of 10,600 by August 2025, and it was still increasing (Table 16.1 and Fig. 16.2). The average number of citations of the science article from 2020 to 2024 was about 500 each year. By August 2025, Rattan Lal had 25 articles with citations of more than 900 each. Consequently, he was ranked no. 1 globally in soil science and agronomy in 2023 and 2024 by Research.com and other academic rankings. In other words, publications were one of his major scientific strengths and a prominent cause of his global recognition. In addition to journal articles, keynote presentations and books (22 authored and 85 edited by 2025) were also an important component of his scientific career.

16.4 Citations of an Article in Science

An article published by Rattan Lal in *Science* in 2004 had a record number of citations, reaching about 10,600 in 2025 (Table 16.1). The average number of citations was more than 500 per year since 2015 and reached 840 in 2024 (Fig. 16.2). This article highlighted the significance of restroring soil organic carbon in agroecosystems with regard to achieving zero hunger and accomplishing many other Sustainable Development Goals (SDGs) of the United Nations Agenda 2030.

Another statisic as an indcator of global reach of the research and education programme was the authorship of refereed journal articles. By the end of 2024, Rattan Lal was the lead author for 315 articles, second author for 440, third author for 174 and other-author status for 216 (Table 16.1). The fact that Rattan Lal's co-authors were from countries around the world indicated the global co-operatioin of his programme (see Chapter 15).

16.5 International Travel

In connection with his research and outreach, specifically making keynote presentations, Rattan Lal had travelled to 103 countries by December 2023 (Table 16.2). The number of countries visited was 22 in Europe, 9 in Africa, 23 in Asia and 13 in South America. The list of 25 highly cited articles, with citations of more than 700 by the end of 2023, is indicative of the interanational significance of the programme and its outreach activities. These citations are reflected in the temporal distribution and the H-Index, which for Google Schoar reached 200 by December 2024 (Table 16.3).

16.6 Science without Borders

Publications (books, journal articles, webinars), large citations and high H-Index were extremely helpful to initiating, promoting and maintaining co-operation. Scientific exchange (visiting scholars, students, short visits and zoom-based webinars) were critically important to these co-operations. While 1000 webinars were presented, these were strictly complementary. All committee assignments and memberships to advisory committees to different organizations were also without any honoraria, and only actual expenses (travel, hotel, food) were paid by the host organization. These co-operative activities were performed as a service to the scientific profession with the moto of 'science without borders' (e.g. political, cultural, religious, ethnical or financial). The dramatic increase in the number of webinars presented internationally for the decade 2016–2025 was influenced by the fact these were free. The number of webinars presented between 12 May and 31 of December (the year the Food Prize was awarded) was 130 to countries around the world. The COVID tragedy had a bright side in the form of a revolution of virtual presentation when no one left the desk and seminars were presented around the globe.

Indeed, global issues of the 21st century (e.g. climate change, soil degradation, water eutrophication, biodiversity loss, air pollution and other processes of environmental degradation, food and nutritional insecurity) do not respect political boundaries or cultural/religious/ethnical lines. Therefore, the Lal Carbon Center must find ways and means and continue its historic path of addressing the current emerging problems of climate change and soil degradation at local, state, national, regional and global scale. The centre must prioritize the key issue of making sustainable agriculture, at all levels, part of the solution towards adaptation and mitigation of climate change and advancing primary SDGs of the UN Agenda 2030.

Table 16.2. Co-operation with overseas universities and institutions.

Europe	
1	Agricultural University of Norway, Ås, Norway (programme on agricultural sustainability in Africa, 1993–)
2	Technical University of Dresden, Germany (2012–)
3	University of Lleida, Spain (2017–)
4	UIMP, Santander, Spain (2018)
5	University of Hohenheim, Stuttgart, Germany (programme on soil erosion in the Andes, Colombia, 1992–1997)
6	Ministry of Environment and Regional Policy, and the Hungarian Academy of Sciences, Budapest (1994–1996)
7	University of Ghent, Belgium, Associate Expert Programme (1976–1987)
8	University of Munich and Gottingen, Associate Expert Programme (1983–1987)
9	University of Reading, UK (1976–1986)
10	Institute of Land Reclamation and Grassland Farming, Falenty, Poland (2000–)
11	IRD/CIRAD France (2002–)
12	Soil Conservation Service of Iceland (2002–2012)
13	University of Iceland (2007–)
14	Institute for Advance Sustainability Studies, Potsdam, Germany (2010–2016)
15	Helmholz Zentrum (GFZ), Potsdam, Germany (2010–2016)
16	Government Office of Science, London, UK (2010–2016)
17	Faculty of Agriculture, Aarhus University, Denmark (2010)
18	University of Bonn, Germany (2010–2016)
19	International Center for Advanced Mediterranean Agronomic Studies (CIHEAM), Valenzano, Italy
20	University of Cordoba, Spain (2008–2018)
21	CIRAD, Montreal, France (1995–)
22	University of Natural Resources and Life Sciences, Vienna (2010–)
AFRICA	
1	Sokoine Univeristy of Agriculture, Morogoro, Tanzania (1980–1985, 1994–1997, 2012–)
2	University of Ibadan, Nigeria (1970–1987)
3	University of Ghana, Legon (1974–1980, 1996–1997)
4	Makerere University, Kampala, Uganda (1988–)
5	Egerton University, Kenya (2012–)
6	Hawassa University, Hawassa, Ethiopia (2010)
7	Addis Ababa University, Ethiopia (2005–)
8	University of Nairobi (2000–)
9	ICARDA, Cairo (2015–)

Continued

Table 16.2. Continued

ASIA	
1	University of Putra, Kuala Lampur (external faculty reviewer, 1995–2000)
2	Tamil Nadu Agricultural University, Coimbatore, India (1998–2012)
3	ANGR Agricultural University, Hyderabad, India (1998–2009)
4	Indian Institute of Technology, Kharagpur, India (1997)
5	Punjab Agricultural University, Ludhiana, India (1998–)
6	Haryana Agricultural University, Hisar, India (1998–)
7	JLNKVV, Jabalpur, India (2001–2012)
8	M.S. Swaminathan Research Foundation (2000–)
9	University of Agriculture, Faisalabad, Pakistan (1997–)
10	Indian Council of Agricultural Research (1995–)
11	Tashkent Institute of Irrigation & Agricultural Mechanics, Uzbekistan (2002–)
12	Bahauddin Zakariya University, Multan, Pakistan (2008–)
13	Ben Gurion University of the Negev, Israel (2008–)
14	Sultan Qaboss University, Oman (2009–)
15	University Putra, Serdang, Malaysia (2000)
16	PMAS Arid Agricultural University, Rawalpindi, Pakistan (2010–)
17	B.Z. University, Multan, Pakistan
18	North-west A & F University, Yangling, Shaanxi, China (2016–)
19	Himachal Pradesh Agricultural University, Palampur, India (2015)
20	North-west A & F University, China (2012–)
21	Indian Institute of Soil Science, Bhopal, India (2010–)
22	Center of Research in Druland Agriculture, Hyderabad, India (2010–)
23	Kathmandu University, Nepal (2014–)
SOUTH AMERICA	
1	University of Palmira, Colombia (1997–2000)
2	University of São Paulo Piracicaba, Brazil (1997–)
3	University of Bahia Blanca, Argentina (1995)
4	Universidad de Buenos Aires, Argentina (2000)
5	University of Audonoma de Santo Domingo, D.R. (2002–2005)
6	Center Para El Desaprollo Agropecuarlo y Forestal, Inc., D.R. (2002–2005)
7	CENBIO, São Paulo, Brazil (2009–)
8	University Federal de Lavras, Brazil (2010–)
9	São Paulo State University, Piracicaba, SP, Brazil (2004–)
10	Universidad e Estadual de Ponta, Grossa, Brazil (2000–)
11	State University of Mato Grasso, Brazil (2016–)
12	Universidad Estadual ed Maringa, PR, Brazil (2010)
13	Universidad Estadual Paulista Julio de Mesquita Filho, Jaboticabal, SP, Brazil (2004)

Table 16.3. Some significant publications with citations of 900 and above (as of August 2025) from Google Scholar).

Publication	Citations	Year
1. Lal, R. (2004) Soil carbon sequestration impacts on global climate change and food security. *Science* 304, 1623–1627.	10553	2004
2. Bronick, C.J. and Lal, R. (2005) Soil structure and management: a review. *Geoderma* 124, 3–22.	5755	2005
3. Lal, R. (2004) Soil carbon sequestration to mitigate climate change. *Geoderma* 123, 1–22.	5241	2004
4. Lal, R. (2005) Forest soils and carbon sequestration. *Forest Ecology and Management* 220, 242–258.	2230	2005
5. Lal, R. (2003) Soil erosion and the global carbon budget. *Environment International* 29, 437–450.	2212	2003
6. Lal, R. (2008) Soils and sustainable agriculture. A review. *Agronomy for Sustainable Development*, 28(1), 57–64.	2119	2009
7. Lal, R. (2001) Soil degradation by erosion. *Land Degradation & Development* 12, 519–539.	2109	2001
8. Lal, R. (2008) Carbon sequestration. *Philosophical Transactions of the Royal Society* B 363, 815–830.	2107	2008
9. Lal, R. (2015) Restoring soil quality to mitigate soil degradation. *Sustainability* 7(5), 5875–5895.	2083	2015
10. Lal, R. (2004) Carbon emission from farm operations. *Environment International* 30, 981–990.	2036	2004
11. Stockmann, U., Adams, M.A., Crawford, J.W., Lal, R. *et al.* (2013) The knowns, known unknowns and unknowns of sequestration of soil organic carbon. *Agriculture, Ecosystems & Environment* 164, 80–99.	1910	2013
12. Liu, R., Lal, R. (2015) Potentials of engineered nanoparticles as fertilizers for increasing agronomic productions. *Science of the Total Environment* 514, 131–139.	1575	2015
13. Paustian, K., Andrén, O., Janzen, H.H., Lal, R., Smith, P. *et al.* (1997) Agricultural soils as a sink to mitigate CO_2 emissions. *Soil Use and Management* 13, 230–244.	1438	1997
14. Lal, R. (2005) World crop residues production and implications of its use as a biofuel. *Environment International* 31, 575–584.	1360	2005
15. Blanco, H. and Lal, R. (2008) Principles of soil conservation and management, Vol. 167169. Springer Science and Business Media, New York.	1354	2008
16. Lal, R. and Shukla, M.K. (2004) *Principles of Soil Physics.* CRC Press, Boca Raton, Florida.	1328	2004
17. Lal, R. (2006) Enhancing crop yields in the developing countries through restoration of the soil organic carbon pool in agricultural lands. *Land Degradation & Development* 17, 197–209.	1321	2006
18. Lal, R. and Stewart, B.A. (date??) Soil degradation. Springer-Verlag.	1216	1990
19. Lal, R. (1998) Soil erosion impact on agronomic productivity and environment quality. *Critical Reviews in Plant Sciences* 17, 319–464.	1098	1998
20. Lal, R. (2016) Soil health and carbon management. *Food and Energy Security* 55(4), 212–222.	1027	2016

Continued

Table 16.3. Continued

Publication	Citations	Year
21. Lal, R. (2002) Soil carbon dynamics in cropland and rangeland. *Environmental Pollution* 116, 353–362.	1020	2002
22. Lal, R. (2004) Carbon sequestration in dryland ecosystem. *Environmental Management* 33, 528–544.	926	2004
23. Shukla, M.K. and Lal, R. (2006) Determining soil quality indicators by factor analysis. *Soil & Tillage Research* 87, 194–204.	921	2006
24. Mulumba, L.N. and Lal, R. (2008) Mulching effects on selected soil physical properties. *Soil and Tillage Research* 98(1), 106–111.	901	2008
25. Lal, R., Reicosky, D.C. and Hanson, J.D. (2007) Evolution of the plow over 10,000 years and the rationale for no-till farming. *Soil and Tillage Research* 93 (1), 1–12.	901	2007

16.7 International Networking with World Leaders

With 18 years at IITA and 40 years at OSU, Rattan Lal had excellent opportunities to establish networks of co-operation with researchers, the private sector (industry), policy-makers and professional societies. These networks proved very helpful to him in learning new ideas, promoting and upscaling his research, organizing conferences, and editing/writing books. Networking is a prudent strategy, which is successful only if it is combined with a solid foundation of research, education and outreach. Some examples of these networks are shown in the photographs below: with Dr Borlaug and Dr Havener of CIMMYT (Fig. 16.6); with IASS in Potsdam, Germany with Dr Klaus Töpfer and Dr Paul Crutzen (Fig. 16.7); and with the team of the One Planet Initiative in Champaign, France (Fig. 16.9). Co-operation with Dr Klaus Töpfer to implement to Global Soil Week at IASS in Potsdam, and with Dr David Malone, Dr Karl-Heinz Feger and Dr Reza Ardakanian to work with UNU-FLORES in Dresden (Fig. 16.8) were professionally rewarding experiences. It is important to note that these co-operative programmes were strictly professional with no financial gain or honoraria involved.

Additionally, Rattan Lal also worked with several UN organizations, especially FAO in Rome and UNCCD in Bonn. Rattan served as a member of the Science Policy Interphase program of UNCCD and worked with Barron Orr for about 4 years. These were all extremely rewarding experiences because of the international scope of the program and their focus on addressing issues of global significance.

16.8 War is a Serious Factor of Soil and Environment Degradation

Rattan Lal has been teaching processes, factors and causes of soil degradation and desertification since early 1970s. Soil degradation is mostly aggravated by anthropogenic land misuse and soil mismanagement. Among all anthropogenic factors, modern war is the most serious factor with long-lasting adverse effects on soil processes, properties and its capacity to generate ecosystem services for humans and nature. War is not only a factor in soil degradation and pollution/contamination, but it also aggravates hunger and malnutrition by disrupting access to food for those who are caught in-between the warring parties. Effects of war on soil degradation, water pollution, air contamination and biodiversity loss can persist for generations to come and the slow process of restoration can occur on a decadal or centennial scale.

Any war has three parties: two fighting with one another and the third party is the land on which the fighting occurs, because both parties believe that 'We Are Right (WAR)'. It is

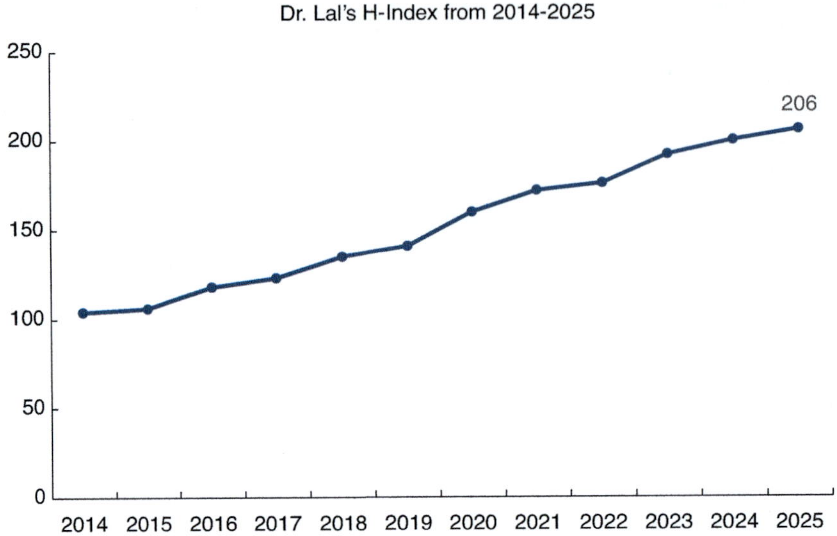

Fig. 16.3. Temporal changes in H-Index from Google Scholar as of 31/7/2025.

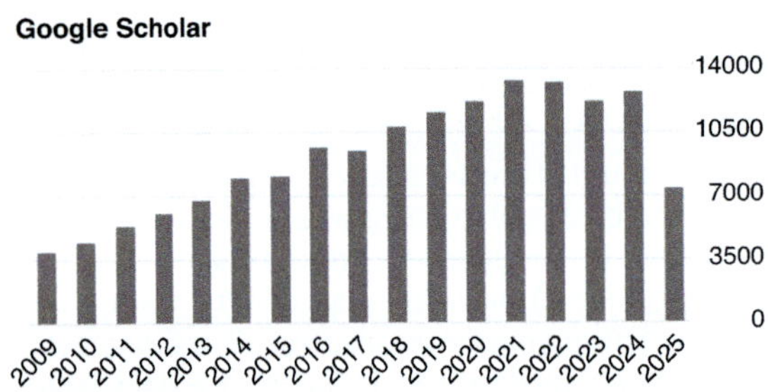

Citation Indices (as of 8/2025): Google Scholar

	All	Since 2020
Citations	181,844	71,200
h-index	206	121
i10-index	1101	736

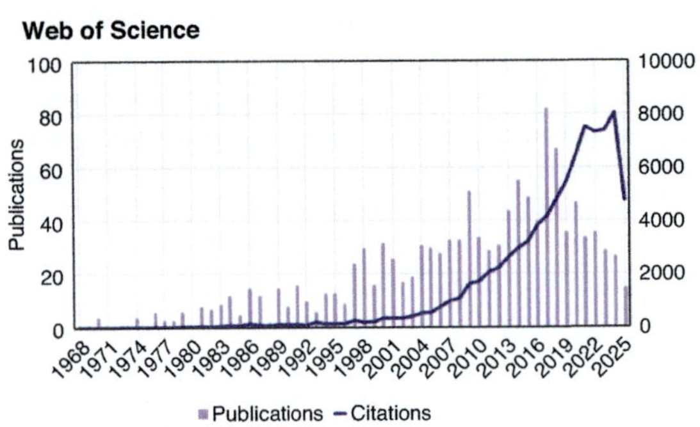

Citations Indices (as of 8/2025): Web of Science

	All
Citations	83,318
h-index	138

Fig. 16.4. Citations per year from Google Scholar (*top*) and publications (purple bars) and citations per year (blue line) from Web of Science (*bottom*).

Fig. 16.5. (*left to right*) Prof. Rattan Lal, Distinguished University Professor of Soil Science, The Ohio State University; Dr Adel El-Sayed Tawfik El-Beltagy, Chairman, International Dryland Center and former Minsiter of Agriculture of egypt; and Dr Kauser Abdulla Malik, Interim Federal Minister of National Food Security and Research of Pakistan. They were co-editors of dryland farming published by CAB International (2024).

Fig. 16.6. Dr Robert Havener, Former Director General of CIMMYT, was awarded an honorary degree at Ohio State in 2003 when Dr Norman Borlaug also accompanied him. (Photo: Courtesy of the Ohio State University)

Fig. 16.7. From 2013 to 2016, Rattan Lal was Senior Science Advisor to Dr Klaus Töpfer (*left*), former Secretary of Environment of Germany, for implementation of the Global Soil Week. At its annual meeting in 2016, Dr Paul Crutzen (*centre*), Nobel Laureate in atmospheric chemistry, was invited to present the keynote address. (Photo: Rattan Lal)

also the land which is the 'silent sufferer' and which no one cares about including the media. Indeed, modern war destroys natural resources and nature is the biggest victim of any war. Even when the peace treaty is signed, neither of the warring parties is held responsible for restoring the degraded soil and restoring injured nature. War is a crime against nature and is not acceptable. There should be a moratorium on war because it the most heinous crime committed by humanity against nature.

16.9 World Peace as a Scientific Issue

Rattan Lal promoted the idea that world peace is indeed a scientific issue. Peace is jeopardized when basic resources (soil, water, vegetation, landscape, biodiversity) are scarce and lead to food and nutritional insecurity. The scarcity of natural resources is also aggravated by anthropogenic degradation, desertification, pollution,

Fig. 16.8. Members of the Scientific Advisory Committee of UNU-FLORES at a meeting in 2018 also met with Dr Reza Andakenian (*4th from left*), the Director of the Centre, Dr. David Malone (*second from right*), and Dr. Karl Heinz Fager (*far right*). (Photo: UNU-FLORES)

scarcity of water and contamination of air. It is a vicious cycle that is often aggravated by political intervention leading to war and civil unrest.

Under these circumstances, the best and long-lasting path to peace is sustainable and judicious management of natural resources, restoration of degraded and desertified soils, conservation of soil and water, reforestation of denuded hills, restoration of desiccated land, purification of contaminated air, improvement of farming, and adoption of nutrient-dense and safe food for each inhabitant of the planet. Access to an adequate amount of safe and nutritious food, the most basic right of each human, is the fundamental basis of long-lasting peace. This is indeed a scientific issue.

Fig. 16.9. During 2020, the One Planet conference was organized in Champagne, France, where the Prime Minister of France was the guest of honour and Rattan Lal was the invited keynote speaker. (Photo: One Planet)

Fig. 16.10. Dr Martin Frick, the Director of the World Food Programme (WFP) Global Office, Berlin (*left*), organized a conference on regenerative agriculture in Germany and invited R.M. Gandhi (*right*), the grandson of Mahatma Gandhi. (Photo: Rattan Lal)

Fig. 16.11. In 2017, Rev. Jesse Jackson and Dr Lal were invited as keynote speakers at the conference of Adapting African Agriculture (AAA) organized by Dr Mohammad Aitkadi and the Prime Minister of Morocco, H.E. Aziz Akhannouch, held in Dahkla, Morocco. The Rev. Jackson was very pleasant and easy to talk with. Most of the discussions we had involved how to promote new scientific ideas among policy makers at national and international levels. (Photo: Adaptation of African Agriculture)

Fig. 16.12. An inter-faith seminar organized in Columbus, Ohio, in 2019 involved a panel discussion on management of soil health by regenerative agriculture. Rattan Lal had the great honour and privilege to be a panellist with former Vice-President Al Gore, which involved exchanges with the audience on the role of religious organizations in promoting regenerative agriculture. (Photo: Rattan Lal)

Fig. 16.13. In 2023, Rattan Lal was invited by the President of Guyana, H.E. Irfaan Ali, and presented a keynote address at the conference chaired by the president. Rattan Lal was very impressed with the public lecturing style of the president, which was informative, eloquent and forceful. It was a pleasure to listen to his ideas on development of a nation. (Photo: Rattan Lal/IICA, San Jose, Costa Rica)

Fig. 16.14. The delegation of The Ohio State University to India led by President Gordon Gee in 2012 had an audience with Her Excellency Pratibha Patil, President of India. (Photo: the Ohio State University)

Fig. 16.15. Mostafa Terrab, Chairman and CEO of the OCP Group, invited Rattan Lal as a keynote speaker at a conference in New York in 2023 to discuss soil health and its management in Africa. (Photo: Rattan Lal)

Fig. 16.16. In 2023, Rattan Lal had the honour and privilege to present his book *The Soil–Human Health Nexus* to N. Modhi, the Prime Minister of India, during his visit to Washington, DC. (Photo: Consul General of India, New York)

Fig. 16.17. A high-level delegation from PAU, led by the Vice-Chancellor and comprising the Registrar and several deans, visited the Ohio State University in 2024. (Photo: Lal Carbon Center)

Bibliography

Davis, N. (2015) Kofi Annan: 'We must challenge climate-change sceptics who deny the facts.' *The Guardian* 3 May. Available at: https://www.theguardian.com/environment/2015/may/03/kofi-annan-interview-climate-change-paris-summit-sceptics (accessed 11 June 2025).

El-Beltagy, A.S., Lal, R. and Malik, K.A. (2024) *Climate Change and Sustainable Agro-ecology in Global Drylands*. CAB International, Wallingford, UK.

Merkel, A. (2014) Lowy Lecture. 17 November. Available at: https://www.theguardian.com/world/2014/nov/17/angela-merkel-pressures-australia-to-reveal-its-greenhouse-gas-targets (accessed 11 June 2025).

17 Basic Laws of Soil Science

'Earth provides enough to satisfy every man's need, but not every man's greed.'
— Mohandas K. Gandhi (1869–1948), quoted in E.F. Schumacher, Small is Beautiful

'We shall require a substantial new manner of thinking if humanity is to survive.'
— Albert Einstein

Abstract

Prominent basic laws of soil science include those developed by Sir Albert Howard since the 1940s and by researchers in the later 20th century and beyond. Important among these are the law of return, the law of the maximum, the law of minima, the One Health concept, soil as a living entity, the rights of soil and ten laws of soil management proposed since the first decade of the 21st century. Adoption of these laws for soils of agro-ecosystems can lead to sustainable management of soil and agriculture for meeting basic necessities of humanity and for transforming agriculture into a part of the solution. Thus, a Soil Health Act can lead to adoption of regenerative practices based on principles of agricultural ecology and carbon farming, which can create another income stream for farmers.

As in other disciplines, soil sciences are also governed by some basic laws. Therefore, prudent management of soil must be based on judicious land use and management as governed by these laws.

Some laws have been known and practised for a long time and others have been developed since the 1960s when rapid advances in soil science happened, especially in sustainable management of soils in diverse eco-regions of the world.

17.1 The Law of Return

Sir Albert Howard narrated the Law of Return, which states that the nutrients extracted from the soil by growing plants must be compensated for by returning their dead remnants to the soil, or fertility loss will occur due to negative nutrient budget. In other words, nutrients removed must be returned in the quantities taken away in order to maintain the balanced ecological state. An example of nutrient mining for a long time is that in agricultural soils of sub-Saharan Africa, where the negative nutrient budget has been estimated at as much as 50 kg/ha of NPK and other elements for a long period of time.

The law also emphasizes the importance of feeding the soil rather than the crop. It is estimated that a square metre of rich healthy soil may contain up to 7 billion living organisms, and that the biomass of living organisms in a healthy soil may weigh as much as 5 Mg (metric tonne) per hectare. Therefore, measurement of crop yield as an indicator of soil health and its functions may not be the best option. Indeed, soil is much more than the medium in which plants grow; it has numerous other functions which depend on the activity and species diversity of soil biota. Therefore, liberal application of biomass carbon is critical to enhancing and sustaining soil health.

It is precisely in this context that Rattan Lal has proposed the concept of CNPK rather than the conventional slogan of NPK. The term C refers to input of biomass-C into the soil which also serves as food for soil biota (macro, meso and micro). As the soil organic matter content

DOI: 10.1079/9781836990154.0017

reaches an optimal level, 2.0–5.0%, corresponding with 1.0–2.0% of SOC content, depending on the climate and soil type, the rate of input of chemical fertilizers and pesticides decreases, because of the improvement in use efficiency of inputs and decrease in leakage of these inputs into the environment.

17.2 One Health

This law or concept is also attributed to Sir Albert Howard (1943). The law states that 'Health of soil, plants, animals, people and environment is one and indivisible'. Thus, degradation of soil health leads to degradation of everything else. Soil degradation has been the cause of extinction of once-thriving civilizations (Mesopotamia, Indus, Mayan, Inca, etc.). This is an important lesson for modern civilization to learn. The One Health concept is in accord with the basic principles of ecology (Carson, 1962; Commoner (1973).

17.3 The Law of Minima

The Liebig Law of Minima states the importance of the balanced application of all nutrients including macro (NPK) and micro (17 needed) for healthy plant growth. The law states that plant growth is determined by the nutrient that is in the most limited supply. In other words, the strength of the chain is determined by its weakest link.

17.3.1 Soil degradation is set in motion by decline in soil organic carbon content

Most complex processes of soil degradation are set in motion by the decline in SOC content to below the critical limit for the site-specific land use and farming/cropping systems. Depletion of SOC content sets in motion a degradative process leading to decline in aggregation and aggregate strength, crusting, water run-off, accelerated soil erosion, drought-flood syndrome, anaerobiosis or waterlogging, and salt imbalance leading to salinization and alkalization.

17.3.2 Living soil

That soil is a living entity is outlined in the book *The Living Soil* (Balfour, 1943). This concept is in accord with that of Aldo Leopold (1949) regarding 'a land ethic', stating:

> Quit thinking about decent land use as solely an economic problem, but examine each question as what is ethically and aesthetically right, as well as what is economically expedient. A thing is right when it tends to preserve the integrity, stability and beauty of biotic community. It is wrong when it tends otherwise.

17.3.3 Soil, soul and spirituality

The idea of soil spirituality is based on the biodynamics basis of management outlined by Steiner during the meetings at Koberwitz and Breslaw in 1924. It emphasizes how a spiritual practice (biodynamics) considers soil as something that is agentic, energetic and alive. Soil should not be considered an inert substrate to be used by its owner as desired.

In addition to these ideas, there are also the 10 tenets of soil management as outlined by Rattan Lal (2009). These laws are outlined below.

17.3.3.1 Law 1

Soil resources are unequally distributed among biomes and geographical regions, and are prone to degradation. Highly productive soils in favourable climates are finite and often located in regions of high population density, and have already been converted to managed ecosystems, e.g. cropland, grazing land and pasture; forest; and energy plantations. Because of a high population and rapid rate of growth, these agronomically productive soils must be protected against urban encroachment and industrialization.

17.3.3.2 Law 2

Most soils are prone to degradation by land misuse and soil mismanagement. Anthropogenic factors leading to soil degradation are driven by desperate situations and helplessness in the case of resource-poor farmers and smaller landholders; and greed, short-sightedness, poor planning and cutting corners for quick economic returns in the case of large-scale and commercial farming enterprises. Therefore, identification and adoption of appropriate land use and site-specific soil/crop/livestock management practices are critical to minimizing the risks of soil degradation. Risks of soil degradation are governed more by how rather than what cropping or farming systems are adopted under specific soil and eco-regional conditions.

17.3.3.3 Law 3

Accelerated soil erosion and decline in soil quality by other degradation processes depend more on 'how rather than on 'what' crops are grown. Productive potential of farming systems can only be realized when implemented in conjunction with restorative and recommended soil and water management practices. Sustainable use of soil depends on the judicious management of both on-site and off-site inputs. Indiscriminate and excessive use of tillage, irrigation and fertilizers can lead to as much, or even more, degradation than no or minimal use of these technologies.

17.3.3.4 Law 4

The rate and susceptibility of soil to degradation increase with increase in mean annual temperature and decrease with a decrease in mean annual precipitation. All other factors remaining the same, soils in hot and arid climates are more prone to degradation and desertification than those in cool and humid eco-regions. However, mismanagement can lead to desertification, even in arctic climates, e.g. Iceland.

17.3.3.5 Law 5

Soil can be a source or sink of greenhouse gases, e.g. CO_2, CH_4 and N_2O, depending on land use and management. Soils are a source of radiatively active gases with extractive farming, which create a negative nutrient budget and degrade soil quality, and a sink with restorative land use and judicious management practices which create positive C and nutrient budgets and conserve soil and water while improving soil structure and tilth.

17.3.3.6 Law 6

Soils are non-renewable over a human timeframe of decadal or generational scales, but are renewable on a geological scale (centennial/millennial). With the increase in human population, projected to be 10 billion by 2100, restoring degraded and desertified soils over a centennial-millennial scale are not an option. Because of heavy demands on finite resources, soils are essentially a non-renewable resource.

17.3.3.7 Law 7

Soil's resilience to natural and anthropogenic perturbations depends on its physical, chemical, biological and ecological processes. Favourable chemical and biological processes enhance resilience only under optimal soil physical properties, e.g. soil structure and tilth; processes, e.g. aeration; water retention and transmission; and edaphological environments, e.g. soil temperature.

17.3.3.8 Law 8

The rate of restoration of the soil organic matter pool is extremely slow, while that of its depletion is often very rapid. In general, restoration occurs on a centennial timescale and depletion on a decadal time cale. The rate of restoration and degradation processes may differ by an order of magnitude.

17.3.3.9 Law 9

Soil structure, similar to an architectural design of a functional building, depends on stability and continuity of macro-, meso- and micropores, which are the sites of physical, chemical and biological processes that support soil's life-support functions. Sustainable management systems, site-specific as these are, enhance stability and continuity of pores and voids over time and under diverse land uses.

17.3.3.10 Law 10

Sustainable management of agricultural ecosystems implies an increasing trend in net primary productivity per unit input of off-farm resources along with improvement in soil quality and ancillary ecosystem services such as increase in the ecosystem C pool, improvement in quality and quantity of renewable freshwater resources, and increase in biodiversity.

Bibliography

Balfour, E. (1943) *The Living Soil*. Faber & Faber, London.
Carson, R. (1962) *Silent Spring*. Hougton Mifflin, Boston, Massachusetts.
Commoner, B. (1973) *The Closing Circle: Confronting an Environmental Crisis*. Jonathan Cape, London.
Howard, A. (1943) *An Agriculture Testament*. Oxford University Press, London.
Lal, R. (2009) Ten tenets of sustainable soil management. *Journal of Soil and Water Conservation* 64(1).
Leopold, A. (1949) *A Sand County Almanac*. Oxford University Press, New York.
Schumacher, E.F. (1973) *Small Is Beautiful: Economics as if People Mattered*. Harper Perennial, New York.
Steiner, R. (1924) Lectures given during the meeting at Koberwitz and Breslaw, Poland.
Whitney, W.D. and Lanman, C.R. (1962) *Atharva-Veda-Saṁhitā*. Motilal Banarsidass.

18 People Are a Mirror Image of the Soil on which They Live

'For we are but a reflection of the land on which we live.'
— *Knut Hamsun,* Growth of the Soil

Abstract

When people are suffering, they pass their miseries to the soil and it reciprocates. The vicious cycle has gripped humanity ever since the dawn of settled agriculture. Many once-thriving civilizations have become extinct because soils were taken for granted and their fragile nature was ignored. People are a mirror image of the soil on which they live. Thus, it is in interests of the present and future civilizations to protect, restore and manage soils for improving human well-being and nature conservancy.

Since childhood, when he lived on a farm in a rural community, Rattan has had a strong interest in the relationship between people and the soil on which they live. He understood from his village life that a healthy soil supports a healthy society and a stable civilization. His place of birth in Karyal, and the village Rajound, where he obtained his high-school diploma, were both located within the site of the extinct Indus Valley civilization. It was the aridization of the climate and the attendant degradation of the soil that led to the extinction of this once mighty and thriving civilization. The story of *Ramayana*, stating that birth of the goddess Sita or Siyya was from the furrow of a plough, was his guiding principle to study the effects of ploughing or its elimination on soil function and quality. Soil is a living entity and it has the capacity to support life. Thus, its degradation has adverse effects on all life dependent on it. When people are poverty-stricken, hungry and miserable, they pass on their miseries to the soil and it reciprocates by returning miseries to the people. Avoiding this vicious cycle through judicious management requires that we acknowledge the rights of the soil.

18.1 Soil and Civilization

A healthy soil supports a healthy society and a stable civilization. This is especially true of predominantly agrarian societies. Throughout human history, thriving civilizations were developed on alluvial flood plains of major rivers such as the Indus, Yangtez, Yellow, Euphrates and Tigris. These civilizations flourished and thrived as long as they looked after their soil resources. At maturity, population of the Indus civilization ranged from 1 to 5 million. However, degradation of soil through neglect and misuse led to the beginning of the end of those once-thriving and mighty civilizations (Montgomery, 2012).

Walking on a footpath from his hometown of Rajound to his farm, Rattan passed through mounds comprising pieces of bricks and pottery, and often wondered who lived here and when, how and why they perished, and even whether these towns were abandoned. Although Rattan did not realize it then, when he was between the ages of 8 and 15, he was walking over the remains of the once-thriving civilization of the Indus Valley. Aridization of the climate and the attendant degradation of the soil, or *vice versa*, led to the extinction of this once mighty and thriving civilization, which was based on the riches of the soils formed in the floodplains of the river Indus.

This was the Bronze Age civilization in the north-western region of South Asia and lasted from 3300 BCE to 1300 BCE, and was at its peak from 2600 BCE to 1900 BCE. Being a contemporary of egypt and Mesopotamia, it was one of the three most ancient civilizations.

DOI: 10.1079/9781836990154.0018

The famous epics of Hinduism, *Ramayana* and *Mahabharata* were composed much later. These epics were composed from the third century BCE to the third century CE.

18.2 Stories from *Ramayana* and *Mahabharata* and Their Influence on a Young Soil Scientist

The festival of Dushera and drama of Ram Leela were played in Rattan Lal's home town of Rajound by actors from the community. The drama of Abhimanue, son of Arjuna, and his assassination during the war of Mahabharata, were among his favourite and heroic acts to watch and admire.

Rattan Lal did not follow the rituals of Hinduism nor of any other religion, but he followed the lessons that he learnt from reading the scriptures (*Ramayana, Gita, Mahabharata*) and from the folk stories that he learnt from his grandmother (bebe).

Among major lessons of *Ramayana* that Rattan learned were: (i) obedience to the parents; (ii) responsibilities of family members towards one another; (iii) problems caused by egotistical and arrogant behaviour; (iv) the mighty power of true love; (v) treating everyone with respect and dignity; (vi) forgive and forget; (vii) effects of bad associations (viii) determination and enthuthiasm; (x) the value of family secrets; and (x) devotion and sacrifice. The triumph of good over evil, and the power of perseverance became Rattan Lal's guiding principles during his school years. The birth of the goddess Sita or Siyya from the furrow of a plough when King Janak of Mathura was advised by his pandits to plough the field and break the long drought was his guiding principle in studying the effects of ploughing and of eliminating it in preparing the seed bed. It was this cultural foundation that led to his professional interest in ploughing, no-till farming and conservation agriculture.

Similarly, the lessons from *Mahabharata* that Rattan Lal learned and cherished were: (i) triumph of good; (ii) sacrifice; (iii) detachment; (iv) forgiveness; (v) loyalty; (vi) duty; (vii) dharma; (viii) leadership; (ix) friendship; and (x) trust. He recited the Gayatri mantra to himself whenever he needed calm and confidence under trying and difficult circumstances. The reciting of the Gayatri mantra had been part of his being throughout his life and especially during the challenging times of his professional career.

18.3 The Gayatri Mantra

The Gāyatri mantra (Sanskrit pronunciation: ga:.jɐ.tri:.mɐn.trɐ.), also known as the Sāvitri mantra (Sanskrit pronunciation: sa:.vi.tri:. mɐn.trɐ.), is a sacred mantra from the *Rig Veda* (Mandala 3.62.10), dedicated to the Vedic deity Savitr It is known as the Mother of the Vedas:

> Om bhūr bhuvaḥ suvaḥ
> tat savitur vareṇyaṃ
> bhargo devasya dhīmahi
> dhiyo yo naḥ pracodayāt

18.3.1 Bhumi and Dharti Mata

The Hindu goddess Bhumi, Bhumi Devi or Mother Earth is also referred to in scriptures by other names such as Bhuvati, Avni, Vasudha, Bhuma-Devi, Dharti, Dharani, Dhaatri, Varahi, Hiranmaya, Hema, Dhanshika, Ira, Urvi, Vasumati, Ela, Kashyapi, Vaishnavi, Vasundhara, Bhudevi, Bhumi or Padmavati (https://www.akkalkotswamisamarth.net/dharti-mata-the-hindu-mother-goddess/ [accessed 13/6/25]).

Bhumi Devi is another avatar of the goddess Laxmi. In Sanskrit, Bhumi is referred to by the term 'Prithvi'. She brings agricultural bounty and is the source of all vegetation. Prajapati is the father of the goddess Bhumi, who is the companion of Varaha, the second avatar of Vishnu. Bhumi is the nurturer and mother of all plants.

Rattan Lal had reverence and respect for Bhumi Devi from his childhood. Any time he travelled to south-eastern countries, he was pleased and proud to witness the roots of this culture. For example, the telex for the Soil Research Institute in Bogor was named Vasundhra or 'The Earth the Beautiful'. The name of the late king of Thailand, on whose birthday the World Soil Day (5 December) is celebrated, is HE Bhumi

Bal Adulya Dej. The name of the airport of Bangkok is Suwran Bhumi or Golden Earth. The Hindu temples in Cambodia are famous tourist attractions.

18.4 The Bhumi–life Nexus

There is a strong connection between life and Bhumi or soil. Charles Kellogg of USDA stated that 'there is no life without soil and no soil without life, they have evolved together' (Kellogg, 1938). Through moderation of properties and processes, and by decomposition of organic matter and release of plant nutrients, billions of organisms in soil create environments which support life. In other words, Bhumi Devi is the only known or proven site in the universe which has the capacity to resurrect or transform death into life. Soil organisms create life-support environments in soil by decomposing organic matter, releasing plant nutrients, transforming and filtering pollutants, forming organo-mineral particles or structural aggregates, which promote aeration and gaseous exchanges, infiltrating and retaining water in soil (the so-called green water that plants can absorb and grow by supporting the process of transpiration and photosynthesis, and creating a favourable medium for plant growth by moderating soil temperature regime).

It is therefore a moral duty and obligation of all generations to protect, restore and sustainably manage soil and hand it over to the coming generation in a better condition than they received it from the previous generation. In 'Laudato Si', the late Pope Francis (2015) emphasized the importance of integral ecology and emphasized the need of seeking comprehensive solutions based on interactions within natural systems themselves and with social systems. Indeed, Pope Francis urged that humanity is not faced with two crises (environmental and social) but only with one complex but also harmonious crisis, which comprises both strongly intersecting components. Therefore, any effective strategy must involve an integrated approach to combating poverty, restoring dignity, and protecting nature. Therefore, the integration of spiritual values from the world's major religions underscores the shared responsibility of humanity to care for the soil as a sacred resource (Lal, 2013, 2023).

18.5 Soil Is a Living Entity

There is a growing understanding of the soil as a living entity and having a strong connection with life. This increases human–soil affections through the development of a sense of shared aliveness, values and ethics. As a living entity, soil has rights (Lal, 2019) and intrinsic value such as that for sequestration of atmospheric carbon (C) (Lal, 2013). Therefore, humans, as stewards of soil, have responsibilities to protect and restore soil's intrinsic value both for human well-being and for nature conservancy. There is a growing need and an urgency to pay more attention to spiritual practices associated with biodynamics and site-specific practices, which enhance soil biota and biotic processes through both spirituality and scientific considerations. It is important to combine technical approaches with spiritual dimensions to protect, restore and sustain soil health. Because soil degradation is also an ethical crisis, spiritual and religious approaches must be combined with scientific innovations to effectively address soil degradation, the most complex and challenging issue of the twenty-first century.

18.6 Rights of Soil, and Soil and Sprituality

Being a living entity, soil has rights to thrive and flourish and be managed properly and judiciously so that it is neither misused nor mismanaged. It is the misuse and mismanagement which is the cause of degradation of soil, pollution of water, contamination of air, and dwindling biodiversity. During the 2020s, an era of the sunset of his career, Rattan Lal wrote articles on the rights of soil (Lal, 2019, 2020, 2024). Reading the article on soil and spirituality, one of his friends sent an e-mail asking, 'Why do these thoughts occur to men when they get old? Rattan replied, 'Because they have nothing better to do'.

18.7 People are a Mirror Image of the Soil

People are a mirror image of the soil they live on. When people are miserable and suffering, they pass on their miseries and suffering to the Bhumi on which they live, and the Bhumi in turn reciprocates. This vicious cycle has gripped humanity at different times throughout human history. The only entry point to break the cycle, which is hard to realize and even harder to enter, is restoration of soil health through its re-carbonization by adopting land-use and soil-management practices which create a positive soil/ecosystem carbon budget. The rate of sequestration of soil organic carbon is low in arid and warm climates compared with those in humid and cold climates. Thus, it is important to implement policies that are pro-nature, pro-agriculture and pro-farmers. These policies must have provisions to reward land managers (farmers, ranchers, foresters) for ecosystem services equivalent to the societal cost of carbon or the real value rather than the one determined by the demand-and-supply of market forces at $50 USD per credit in the 2020's.

Bibliography

Francis, Pope (2015) Laudato Si'; encyclical letter of the Holy Father Francis on care for our common home. Available at: https://www.vatican.va/content/francesco/en/encyclicals/documents/papa-francesco_20150524_enciclica-laudato-si.html (accessed 13 June 2025).

Hamsun, K. (1917) *Growth of the soil*. Penguin.

Kellogg, C.E. (1938) *USDA Yearbook of Agriculture*. U.S. Dept. of Agriculture, Washington, DC.

Lal, R. (2013) Soil and Sanskriti. *Journal of the Indian Society of Soil Science* 61(4), 267–274.

Lal, R. (2019) Rights-of-soil. *Journal of Soil Water Conservation* 74(4), 81A–86A. DOI: 10.2489/jswc.74.4.81A.

Lal, R. (2020) *The Soil-Human Health Nexus*. CRC Press, Boca Raton, Florida.

Lal, R. (2021) Feeding the world and returning half of the agricultural land back to nature. *Journal of Soil and Water Conservation* 76(4), 75A–78A. DOI: 10.2489/jswc.2021.0607A.

Lal, R. (2022) Sustaining soil for advancing peace: The world is one family. *Journal of Soil and Water Conservation* 77(3), 43A–47A. DOI: 10.2489/jswc.2022.0411A.

Lal, R. (2023) Farming systems to return land for nature: It's all about soil health and re-carbonization of the terrestrial biosphere. *Farming Systems* 1(1), 100002. DOI: 10.1016/j.farmsys.2023.100002.

Lal, R. (2024) Soil, soul, spirituality and stewardship. *Journal of Soil and Water Conservation* 79(1), 10A–14A. DOI: 10.2489/jswc.2024.1129A.

Montgomery, D.R. (2012) *Dirt: The Erosion of Civilizations*. University of California Press Books. 296 pp.

19 Notable Quotes by Rattan Lal

'You are the eternal, the timeless, the formless' (interpretation of the Sanskrit phrase 'Tat Tvam Asi'; original: तत् त्वम् असि; literal: 'That Thou Art' or 'You Are That')
— Chandogya Upanishad *6.8.7 and 6.12.3 of the* Sama Veda ॐ

'To forget how to dig the earth and to tend the soil is to forget ourselves.'
— *Mohandas Gandhi, in Back to the village', in* The Mind of Mahatma Gandhi

Abstract
Rattan Lal's beliefs in soil and the necessity of its sustainable management for strengthening of life-support processes are featured in his publications, presentations and classroom teaching. These beliefs in relation to soil health, its effect on humanity and the environment, and on human–soil interaction are the basis of notable quotes cited herein. These quotes are a synthesis of complex processes in simple and easy-to-understand language.

19.1 Quotes from Webinars

Rattan Lal made more than 1000 presentations as an invited keynote speaker throughout the world (see Chapter 16). During these presentations, along with more than 600 book chapters and 1150 refereed journal articles, he made several comments about soil and its uses and abuses in relation to the impact on human well-being, and global peace and stability. These are sometimes used as quotations. Some examples of these are given below.

19.1.1 Soil degradation

Indiscriminate use of inputs based on fossil fuel has aggravated soil degradation by numerous processes.

Soil is like a bank account: it is not possible to withdraw more than what is put in without degrading its quality and capacity to generate ecosystem services.

Natural processes of soil formation are transformed into degradation processes by land misuse and soil mismanagement.

Soil degradation can happen by an isolated activity but it is restored on a decadal or generational scale.

19.1.2 Soil and human

Soil flourishes without human intervention; humans cannot survive without soil.

Human health and well-being are strongly influenced by soil health.

Human actions can restore or degrade soil health.

Soil degradation is caused by human greed, short-sightedness and activities to get rich quick.

Soil responds to human deeds: good by producing more, and bad by aggravating starvation and miseries.

Humans are a mirror image of the soil on which they depend: they thrive on a healthy soil and perish on a degraded and depleted one.

19.1.3 Soil and climate change

Soil mismanagement makes it a source and its restoration a sink of greenhouse gases.

Fossil fuel has substitutes but ecosystem services derived from soil are unique.

Converting soil from natural to managed ecosystems aggravates climate change; returning

DOI: 10.1079/9781836990154.0019

soil to nature mitigates anthropogenic climate change.

Soil and climate are intricately interconnected and affect one another's properties and processes.

Soil is the delicate link between humans and climate.

Humans are the parasites that have caused fever to soil.

19.1.4 Soil and human civilization

Taking soil for granted led to the extinction of numerous once-thriving civilizations.

Regeneration of degraded soils is essential to minimize risks of extinction of the carbon civilization.

Treating soil like dirt aggravates the risks of extinction of all biota that depend on soil, including humans.

Humans come from, and pay their debt by returning to, soil and become a part of its life-support processes.

19.1.5 The Soil-Peace Nexus

Healthy soil is essential to world peace.

Soil flourishes and restores when the world is at peace and it is degraded during war and civil strife.

A war involves three parties: two nations at war, and the soil on which the war is pursued.

19.1.6 Soil and peace

Soil health and peace are mutually enhancing processes.

Peace is jeopardized with an onset of soil degradation which threatens to weaken ecosystem services.

Protection and sustainable management of soil health are essential prerequisites to a long-lasting peace.

There can be no peace without healthy and productive soil.

The ferocious fire that burns in the pit of an empty stomach can only be quenched by a loaf of bread made from grains grown on a healthy soil.

There can be no peace among people with empty stomachs.

If you wish for peace, begin with restoration of soil health and its capacity to provide critical ecosystem services.

Soil health is the foundation of peace and harmony.

Peace is threatened when soil is degraded.

19.1.7 Soil and war

Degradation and pollution of soil are among the most severe casualties of war.

Healing of soil and nature is a slow process and occurs over a decadal/generational scale after the war is over.

War destroys soil health by vehicular compaction, cratering, pollution with heavy metals, destruction of vegetation and biodiversity, and creation of disservices by degradation processes.

Warring parties neither care about the soil nor the nature over which they fight.

When the peace treaty is signed and armies return to their barracks, degraded soil and disrupted ecosystem processes are ignored and left to fend for themselves.

War is a heinous crime against soil and Mother Nature.

19.1.8 Anthropogenic soil degradation

When people are suffering and desperate they pass on their miseries to the soil.

Soil degradation and poverty are mutually reinforcing processes.

Soil degradation leads to under-nutrition and malnutrition of people.

Soil degradation is the primary cause of human malnutrition.

Restoration of soil health is the first step in improving human health and well-being.

Soil degradation processes occur on an annual timescale but those of restoration on decadal, generational, and centennial scales.

19.1.9 Soil and water

Soil is the largest reservoir of fresh water.

Dirt is the best filter to clean dirty water.

Water pollution occurs when it moves over the soil surface, and purification when it passes through the soil.

Sahara comprises of the sand floating over the fossil water.

Soil, with its water and nutrient reserves, is the foundation of all terrestrial life.

19.1.10 Soil and civilization

Restoration of soil health is followed by prosperity, arts and culture.

Civilizations thrive and flourish when soil health is sustained.

Deliberate degradation of soil health will lead to vanishing of a civilization.

Historic civilizations were destroyed by intentional pollution/degradation of their soil.

Civilizations perish when their soils go extinct.

19.1.11 Soil and prosperity

Healthy soil and prosperity co-exist.

Loss of prosperity is a symptom of soil degradation and *vice versa*.

Prosperity follows closely behind the will to restore soil.

19.1.12 Soil degradation and international boundaries

Processes of soil degradation do not respect national or political boundaries.

Soil and environmental degradation are common enemies of two warring nations.

A long-lasting peace will follow among two warring nations when they focus on mitigation of the soil degradation processes.

19.1.13 Soil health and political willpower

Political willpower is critical to restoration and sustainable management of soil health.

Sustainable management of soil health requires pro-soil, pro-nature, pro-agriculture, and pro-farmer policies.

Visionary leadership implies policies which are pro soil health.

A political platform devoid of a soil-health agenda is shallow and self-destructive.

19.1.14 Soil and education

Soil education must be integral to curricula beginning from kindergarten through primary, middle, high school, and college education.

Soil education is important to attract the best and the brightest pedagogy.

Education without soil science is like an empty vessel which makes much noise but has no substance.

19.1.15 The food–energy–water–soil nexus

Soil health is critical to adequate supply of food, energy and water.

The interconnectivity between food, energy and water is described best by the FEWS nexus.

Safe, healthy and nutritious food comes from the strengthening of the FEWS nexus.

19.1.16 Soil and global food security

In addition to improved seed, the 21st-century Green Revolution must be based on soil health and its management.

The seed-centric Green Revolution of the 1960s by-passed sub-Saharan Africa because of ignoring the severe problem of soil degradation.

Long-lasting food and nutrition security follow closely behind comprehensive initiatives to restore health of degraded soils.

South-south co-operation is critical to achieving food security in sub-Saharan Africa.

The pace of a soil-centric approach to achieving food security is slow but pure and long-lasting.

19.1.17 Soil health and the private sector

An active involvement of the private sector is critically important to restoring and sustaining the health of degraded soils of agro-ecosystems.

The private sector can promote sustainable management of soil health through supply of essential inputs.

Research and education on soil health are advanced through partnership between the private and public sectors.

19.1.18 Soil health and sustainable development

Sustainable development is achieved through sustainable management of soil health.

Sustainable development and soil health are mutually enhancing processes.

Good soil health is critical to sustainable development.

19.1.19 Soil and global challenges

Hunger is a human-made problem, and its solution lies in not taking food and soil for granted.

Soil must be appreciated for the good things that it provides.

Poverty among resource-poor farmers is a solvable problem through science-based soil management.

Judicious land use and sustainable soil management can address some of humanity's greatest challenges of the 21st century.

Sustainable soil management is a pathway to a positive future for global agriculture.

Protecting and restoring soil health must receive unprecedented attention to address unprecedented issues of the growing and increasingly affluent world population.

Sustainable soil management and restoration of soil health are the foundation of regenerative agriculture based on basic principles of agroecology.

Recarbonization of soils of agro-ecosystems is critical to optimization of soil health and enhancing use efficiency of inputs.

19.1.20 Soil health for producing more from less

Managing soil resources is critical to enhancing use efficiency of inputs and producing more from less.

Enhancing soil health is critical to winning the fight for a better planet.

Passion for improving the environment is spurred by the need for producing more from less and for restoring soil health.

Sustainable development can only be achieved by protecting, restoring and recarbonizing soils of agro-ecosystems.

Enhancing soil health is essential to scaling up a system-based regenerative agriculture.

Re-carbonization of depleted and degraded soils of agro-ecosystems is an essential prerequisite to environmentally friendly, economically viable and socially acceptable agriculture.

Adopting environmentally compatible, socially responsible and economically profitable agriculture begins with re-carbonization of the depleted and degraded soils.

Soil health-promoting practices begin with sequestration of carbon and optimization of pedological processes.

Sustainable food systems must spare soil, water and other resources for nature.

Sustainability of crop farming depends on protection and restoration of soil carbon stock.

Green water and plant nutrients can be saved by restoring soil health.

Advancing sustainable agriculture can only be achieved by restoring and judiciously managing soil health.

Optimizing soil organic carbon stock in the root zone is critical to addressing both hunger and hidden hunger.

Reference

Gandhi, M. (1946) *The Mind of Mahatma Gandhi*. Oxford University Press.

20 Spiritualism in Agriculture and Mystics of Soil among Ancient Cultures

'We are the Earth. There is no environment out there. The Earth is our extended collective body.'

— *Kyle Pearce*

Abstract

Most ancient cultures and tribes believed in deities (gods and goddesses) with divine powers to address specific problems of agriculture. Some of these are considered responsible for maintaining soil fertility, good harvest, seed germination and crop stand, and rain to mitigate drought. With advances in scientific development and understanding of the basic properties and processes of soil, traditional gods and goddesses have been replaced by some key disciplinary terms used in modern scientific literature such as pedology, agronomy, agroecology, biotechnology, plant physiology, climatology, etc. Adoption of scientific innovations has alleviated some of the constraints faced by farmers, and has improved agronomic productivity since the middle of the 20th century.

20.1 Soil and Spirituality

Ever since the beginning of agriculture, humans have looked up to divine powers for advice and help with their agricultural operations such as planting, pest infestation, weather, large harvest, protection of livestock and of the land against natural and human-made calamities. Droughts have been the biggest threat to food security throughout human history and ever since the dawn of settled agriculture about 10–12 millennia ago. Thus, most ancient cultures and tribes believed in deities (gods and goddesses) with divine powers to address specific problems of agriculture. Notable among these are those believed responsible for maintaining soil fertility, good harvest, seed germination and crop stand, and rain to mitigate drought. Examples of such deities worshipped by traditional cultures and tribes are presented in Table 20.1.

All terrestrial life depends on soil fertility. Mother Earth is the most widespread concept among traditional cultures across the world. Other deities include those for good harvest, rainfall, soil fertility, grains, agriculture and crops. Such beliefs are widespread across cultures and those depicted herein are merely some examples.

Major religions of the world have specific prayers for Mother Earth and the need for safeguarding and judicious management of soil and natural resources. Ploughing and soil are prominent parts of several ancient cultures and civilizations. It is thus pertinent that religious organizations emphasize these in each of the religions. This is an important common point in scriptures of all religions and is the basis of unity among different beliefs and philosophies (Table 20.2).

All contemporary religions of the world revere and worship Mother Earth (Bhumi Devi) as a common denominator (Table 20.2; Lal, 2013). The growing and complex problem of soil degradation, increasing with increase in population and its affluent lifestyle, may need more than just biophysical or technological solutions. Because soil and environmental degradation are spiritual crises, the problems cannot be solved only by scientific, technological, economic or political approaches. Indeed, a broader approach, also encompassing religious and spiritual beliefs, may be necessary to address

DOI: 10.1079/9781836990154.0020

Table 20.1. Some examples of cultural philosophy that land gifts life. (Adapted from Lal, 2007a; Evans, 2021)

Deity	Country/culture	God/goddess of:
Atabey (F)	Pre-Colombia America	Mother Earth
Bhu Devi (F)	India	Earth goddess
Ceres (F)	Ancient Rome	Agriculture, crops, fertile land, grains
Demeter (F)	Ancient Greece	Grains, agriculture and fertile soil
Devi Sri (F)	Bali (Indonesia)	Earth, fertility goddess
Dherra (F)	India	Mother Earth
Hequet (F)	Egypt	Flooding of Nile, germination of grain crops; symbolized as a frog
Indra (M)	India	Rain god
Otikan (F)	Turkey	Mother Earth
Pachamama (F)	Andes-Inca	Earth Mother, planting, and harvesting
Siya/Sita (F)	India	The furrow
Terra Matre (F)	Italy	Mother Earth
Tudi Gong (M)	China	Harvest

Table 20.2. Soil, soul, stewardship and spirituality among different religions of the world. (Adapted from Lal, 2013, 2024)

Spiritual beliefs in soil and environment	
Judaism	The word 'homo' (man) is derived from the Latin word 'humus' or the decomposed organic matter in soil, which is the essence of all terrestrial life. The Hebrew phrase 'Tikkun Olam' means 'repairing or restoring the world'.
Hinduism	The human body is made of 'Kshiti (soil), Jal (water), Pawak (energy), Gagan (sky/space), Sameere (air)' (*Prasna Upanishad*).
Sikhism	Pau'ṇ gurū pāṇī pitā mātā dharaṯ mahaṯ. Dinas rāṯ du ė ḏā ī ḏā i ā kẖėlai sagal jagaṯ. (Gurbani)
Buddhism	'One should not break even the branch of a tree that has given one shelter' (*Petavatthu* II, 9, 3).
Christianity	The word 'Adam' (man) is derived from the Hebrew word 'adama' meaning 'earth' or 'soil.'
Greek	The daughter of the Earth goddess 'Gaea' named Themis (goddess of Law), and her descendent Demeter was the goddess of agriculture and fertility.
Roman	The Earth goddess (Tellus) was related to the goddess of fertility and harvest (Ceres), and named 'Mater Terra'.
Islam	'He created the man of clay like the potters' (*Suhrah Al Rhman*, verse 14) 'We made from water every living thing' (*Qur'an* 25:54) 'Do not overuse water even if you are on a running river' (Prophet Mohammad).
Khalil Gibran	Trees are poems (rubbiat) that earth writes upon the sky. We fell them down and then turned them into paper, so that we may record our emptiness.

this complex and aggravating issue of global significance. However, spiritual knowledge and practices do not fit comfortably with technoscientific and productivity accounts of soil health and its management. Therefore, there is a need for stronger and continued dialogue among researchers, land managers and spiritual leaders to develop a common approach. Despite the growing realization about the importance of spiritual elements to sustainable management of the goddess Bhumi, there is a tendency to neglect discussion of these aspects among different stakeholders in this finite but a precious resource.

Increase in human population is the major driver of food and nutritional insecurity. It took 200,000 years for the human population to reach 1 billion in 1803. It took a further 123 years to reach 2 billion, 34 years more to reach 3 billion, 15 to reach 4 billion and only 12 more years to reach an additional billion, reaching 8 billion by 2023 and 9 billion by 2037 (Connection, 2024). As much as 96% of the global population growth since the 7 billion milestone has occurred in developing countries. This rapid global expansion of population has caused extinction of several species (Smithsonian, 2024). Therefore, in the long term, a viable option is to return some agricultural land back to nature through the soil-centric approach being proposed in this book.

20.2 Modern Gods and Goddesses of Soil and Agriculture

Ancient gods and goddesses (Table 20.1) have been replaced by some key disciplinary terms used in modern scientific literature. These are: pedology, agronomy, agroecology, biotechnology, plant physiology, climatology, etc. Indeed, adoption of scientific innovations has alleviated some of the constraints faced by farmers. Thus, agronomic productivity has drastically improved since the middle of the 20th century. Indeed, agriculture has a bright future ahead. More changes will occur between 2020 and 2050 than have occurred since the beginning of agriculture over 12 millennia ago through building scientific innovations on top of traditional knowledge and wisdom.

Bibliography

Evans, M. (2021) Soils: The Incredible Story of What Keeps the Earth and Us Healthy. Murdoch Books, Murdoch, Australia. ISBN: 9781922351418.

Handelsman, J. (2021) *A world without Soils: The Past, Present and Precarious Future of the Earth Beneath our Feet*. Yale University Press, New Haven, Connecticut. ISBN: 9780300256406.

Jenik, J. and Hall, J.B. (1966) The ecological effects of the Harmattan wind in the Djebobo Massif (Togo Mountains, Ghana). *Journal of Ecology* 54(3), 767–779.

Lal, R. (2007a) Soil science and the carbon civilization. *Soil Science Society of America Journal* 71, 1095–1108.

Lal, R. (2007b) World soils and global issues. *Soil Tillage Research* 97(1), 1–4.

Lal, R. (2013) Soil and Sanskriti. *Journal of the Indian Society of Soil Science* 61, 267–274.

Lal, R. (2024) Soil, soul, stewardship and spirituality. *Journal of Soil and Water Conservation* 79(1), 10A–14A. DOI: 10.2489/jswc.2024.1129A.

Pearce, K. (2023) *100+ Spiritual Nature Quotes on Beauty, Wisdom & Adventure*. Available at: https://www.mindfulecotourism.com/spiritual-nature-quotes/ (accessed 13 June 2025).

Population Connection (2024) World population milestones through history. Available at: https://populationconnection.org/learn/population-milestones/ (accessed 13 June 2025).

Smithsonian (2024) Extinction of plants and animals. National Museum of Natural History, Washington, DC. Available at: https://naturalhistory.si.edu/education/teaching-resources/paleontology/extinction-over-time (accessed 13 June 2025).

21 Hunger Never Remains Hungry: It Feeds on the Flesh and Blood of Victims

'Hunger is not an issue of charity. It is an issue of justice.'
— *Jacques Diouf, Director General, Food and Agricultural Organization of the United Nations*

'A hungry man can't see right or wrong. He just sees food.'
— *Pearl S. Buck*

Abstract

Rattan Lal's grandmother reminded him that food (*anna* in Sanskrit) is God and should neither be disrespected nor refused. Food must be accepted with prayers and gratitude, and eaten with utter respect and absolutely no waste. She also advised that hunger is a very greedy and revengeful enemy of humanity; it never stays hungry, and it feeds on the flesh and blood of the unfortunate, those who do not have access to an adequate amount of healthy, nutritious and safe food. This philosophy is the basis of Rattan's thinking about treating food as God, its consumption as a religious ritual, and its waste as a sinful act.

21.1 Mahabharata

During his childhood, Rattan listened stories from the *Mahabharata*. It relates the stories of the escape of the Pandava brothers and Drupadi from the flammable Lac palace via an underground tunnel; the donation by warrior Karan of his golden teeth to a saint who called on him for alms when he was mortally wounded and lay helplessly on the battleground; the receding of the river Yamuna when Vasudev was taking her baby Krishna to safety following the attempt on Krishna's life by the maternal uncle Kans; and many others. These stories were the cultural foundations of Rattan's thinking about soil and its management, and the need for treating food as God, its consumption as religious ritual and its waste as a sinful act.

Whenever Rattan Lal got annoyed about some trivial things, he would get upset and say to his grandmother that he would not eat food. His grandmother always reminded him that food is God and should neither be disrespected nor refused. Food must be accepted with prayer and gratitude, and eaten with utter respect and absolutely no waste. She would also warn Rattan that hunger is a very greedy and harmful enemy of humanity. It never stays hungry it and feeds on the flesh and blood of those who are unfortunate and who do not have enough healthy, nutritious and safe food to eat.

During his career as a soil scientist, Rattan Lal learned from the teachings of Gandhi and many other philosophers that, indeed, food is God. He believed in the teaching of Gandhi that to a hungry man, God dare not appear in any form other than a loaf of bread. In his own lectures on hunger and malnutrition, Rattan has promoted the concept from *Upanishad* that healthy and nutritious food is medicine: 'When food is right, medicine is not needed, and when food is wrong, medicine is of no use' (*Prasanna Upanishad*).

DOI: 10.1079/9781836990154.0021

21.2 Global Hunger and Malnutrition

There are 733 million people prone to hunger and another 2 billion to malnutrition in 2025. The number of people prone to under-nutrition was 590 million in December 2019, prior to the COVID pandemic. The number is projected to increase by 2030 because of the conflict and political instability leading to internal displacement and increase in the number of global refugees. The problem is further aggravated by climate change and increase in intensity and severity of drought (see Fig. 5.2).

In addition to under-nutrition affecting 733 million, there are more than 2 billion people prone to malnutrition or hidden hunger. Malnutrition is caused by deficiency of essential micronutrients and protein. Essential micronutrients for good human health and well-being number 17, and prevalent among deficient nutrients are Fe, Zn, Mo, Cu, I, B, Mn and Se. Degradation of soil and depletion of soil fertility are among major causes of malnutrition in developing countries of sub-Saharan Africa, South Asia, the Caribbean, the Andean region, etc.

With the aggravating problem of hunger and hidden hunger, or malnutrition, in the world, the importance of safe and healthy food can never be overstated.

21.3 Transforming the Food System

The amount of food produced (more than 3 Gt of cereal grains per year) is adequate to feed the present population of 8.2 billion in 2025 and a future population projected at about 10.2 billion by 2080 (UN, 2024). Important strategies for feeding the world population involve drastic transformation of the world food system. While agriculture (growing of food on land and in aquatic ecosystems) is the first step, processing and value addition, packaging, transport or distribution, storage, sale, cooking, and disposing of waste are among a series of steps involving food systems. While agriculture contributes 12–15% of the anthropogenic greenhouse gases (GHGs), food systems as a whole contribute about 30% of anthropogenic emissions.

Food systems are in need of drastic transformation because they require a vast amount of resources but are utterly inefficient and ineffective in feeding the increasing and progressively affluent world population. For example, 40% of the Earth's surface (more than 5.5 billion ha) is used for agriculture. As much as 77% of agricultural land (almost 4 billion ha) is allocated to raising animals. In addition, 70% of freshwater withdrawal is used for irrigation and 30–35% of all GHGs are contributed by food systems.

Yet, 733 million (1 in 11 persons) are food-insecure, and 2–3 in eight are prone to hunger and malnutrition. The current food systems are vulnerable to climate change and related extreme events such as the drought-flood syndrome, heatwave, and increase in incidence of pests and pathogens. Food waste, estimated at 30–50% or more is a serious problem in both developed and developing countries.

As the COVID-19 pandemic demonstrated, food systems are extremely vulnerable to disruptions in both production and supply chains. The problem has been aggravated by global conflict and instability. Food systems must be resilient against biotic and abiotic stresses and resilient to civil strife and political instability. In this regard, local food production systems must be strengthened such as urban farming and home gardens.

21.4 Feeding the World Population by 2050

The world produces enough food to feed 10 billion people. However, the food systems must be transformed with regard to addressing the following issues which require immediate attention:

21.4.1 Reducing food waste

This is a global issue, and 30–50% of all food produced is wasted in both developed and developing countries. In developed countries (e.g. North America, western Europe, East Asia, Australia), food is wasted between the grocery store and the kitchen table; and in developing countries (e.g. South Asia, sub-Saharan Africa, the Caribbean), food is wasted between the farm and the grocery store. If food waste was a country, it would be the third largest emitter of GHGs.

21.4.2 Increasing food access

Access to food is seriously affected by global conflicts and instability. There is a strong nexus between soil health and global peace.

21.4.3 Realizing that famines are man-made tragedies

In the current era of bountiful agricultural production since the 1960s when the rate of agricultural growth has exceeded the rate of population growth, famines and mass starvation are man-made tragedies. We must make famine and mass starvation politically intolerable, morally toxic, ethically unthinkable and humanely unacceptable.

21.4.4 Mitigating the drought-flood syndrome

The drought-flood syndrome, aggravated by the climate change and related extreme events with severe adverse effects on agriculture, is also a man-made tragedy. Drought-flood problems are caused by land misuse and soil mismanagement. Take the case of the Himalayas. Countries of South Asia (Pakistan, India, Nepal, Bangladesh) suffer from flood in the monsoon season and drought during the dry season. The Chirapoonji region in north-eastern India, receives the world's highest annual rainfall of 500 in (1250 cm per annum). Yet, it suffers from drought when it does not rain for just one month.

It is the denuding of the Shivalik Hills from Afghanistan in the west to Cambodia in the east which is the cause of floods in the monsoon season. Since most water is lost as run-off, causing havoc by floods during the monsoon, little water is stored within the watershed (surface pond, groundwater recharge or soil-water storage). There is therefore a prevalence of drought when it does not rain. Precipitation (rain or snow) must be stored where it falls as green water, recharge of the groundwater, and river flow.

21.4.5 Improving food distribution

Food distribution must be improved in food-scarce areas by improving road and railway networks and secure storage faciliites. Lack of storage facilities and no efforts to enhance the shelf life by value addition and packaging facilities are the causes of food shortages.

21.4.6 Increasing use of plant-based protein

Animal-based protein is a good source of protein and plant-based protein (grain legumes and pulses) is a viable alternative, often at low cost.

21.4.7 Increasing local food production

Home gardening, community gardens and other local sources of food production must be strengthened. The slogan 'One Home One Garden' must be promoted.

21.4.8 Reycling of food waste

Any food waste must be recycled to make compost and reused on home gardens to minimize transport to landfill and reduce the need for chemical fertilizers.

21.4.9 Narrowing the yield gap and vertical intensification

There exists a large yield gap between what is the national average yield of food staples and what can be achieved. The yield gap must be narrowed by adoption of proven best-management practices.

21.4.10 Eco-intensification and producing more from less

Food production must be increased by improving productivity per unit of land and resource use, and by adopting the principle of producing more from less with the objective of returning some land and water back to nature through eco-intensification.

21.4.11 Focusing on traditional or orphan foods

The Green Revolution of the 1960s was cereal-centric and focused mostly on wheat and rice. Traditional food crops including cereals such as sorghum, millet, barley, oats, rye and teff were excluded. It also ignored pulses such as chick-pea, pigeon pea, mung bean, urd bean, lentil etc. These crops are more nutritious than wheat and rice. Thus, future emphasis should be on traditional crops grown on healthy soils by process of agroecology and regenerative agriculture.

Bibliography

UN (2024) Three billion people globally impacted by land degradation. 2 Dec 2024. https://news.un.org/en/story/2024/12/1157651 (accessed 4 September 2025).

22 Yeh Duniya (This World)

'The dunya is designed to break your heart, so crush it.'
— Sheikh Hamza Yusuf

'If you allow dunya to own your heart, like the ocean that owns the boat, it will take over. You will sink down to the depths of the sea. You will touch the ocean floor.'
— Yasmin Mogahed

'Duniya ke maije hargiz bhi kam nan honge, Churcha yahi rahe gi aphsos hum nan honge.'
('The wonders and good things of the world will never cease. The dialogue about the wonders of this world will continue for ever. Alas, I will not be here to experience these new wonders.'
— anonymous (but the couplet was often mentioned to Rattan Lal by his father during his childhood)

'Iss duniya ke gham, jane kab honge kaam'
— Nusrat Fateh Ali Khan

'...umr-e-darāz maañg ke laa. ***ī*** *th***ī** *chaar din do aarz***ū** *meñ kaT ga.e do intiz***ā***r meñ'*
('...from the gifted life of four days, two passed by in wishing and two in waiting')
— attributed to Bahadur Shah Jaffar, but may have been written by Seemab Akbarabadi

Abstract

Between 1970 and 2025, Rattan Lal travelled to some 103 countries. Thus, he learnt a lot about native food, culture, traditions and agricultural production systems in relation to site specificity of soil, climate and farming systems. The use of the word 'duniya' for COP29 in Baku reminded him of his favourite songs from some Bollywood movies based on this Turkish word meaning 'the world'. Rattan wrote the final chapter of his autobiography on the basis of his experiences – some good and some not so good – that he experienced during his travels.

22.1 Lifetime Experiences

Life is full of ups and downs, joys and sorrows, riches and poverty, good health and sickness, light and darkness, and peace and war. The strategy is to be calm and prudent in evaluating each situation and responding in a way that is appropriate to the situation. As we know, nothing is permanent, nothing is impossible. Some quotes on nothingness are prepared by Kuhn (2013) and Keats (n.d.). However, 12 concepts of nothingness are listed below (adapted from Lal, 2013).

22.1.1 Ecosystem services and the law of nothingness

1. Nothing is appropriated: there are always trade-offs (give and take).
2. Nothing is permanent: everything is in a dynamic equilibrium and a transient state.
3. Nothing is absolute: all processes, properties and values are relative to a baseline.
4. Nothing is a panacea: there is no silver bullet, there is a multitude/menu of options.
5. Nothing is universal: soil-/site-/region-specificity is an important consideration which cannot be overlooked.

DOI: 10.1079/9781836990154.0022

6. Nothing tangible is free: undervaluing a commodity leads to 'tragedy of the commons'. As Berry Commoner (1971) stated, 'There is no such thing as a free lunch'.
7. Nothing is empty (vacuum) in nature: all space is occupied, pores in solid rock contain water or air and injecting something (liquid CO_2), as in fracking, can create shockwaves.
8. Nothing is a given or for granted: it is the judicious use and management which produce goods and services.
9. Nothing is a waste: everything in nature has a use.
10. Nothing comes from nothing (*Ex nihilo nihil fit*): this means 'out of nothing, nothing is produced' or 'nothing comes from nothing'.
11. In nature, nothing exists alone (Muir, 1869; Gussow, 2024).
12. A thing of beauty will never pass into nothingness (Keats, n.d.).

These concepts of nothingness in the context of soil science are described below.

22.1.2 Soil's law of nothingness

1. Nothing is free: trade-offs
2. Nothing is permanent: time-dependence
3. Nothing is absolute: relative magnitude
4. Nothing is a panacea: no silver bullet
5. Nothing is universally applicable: site-specificity
6. Nothing comes from nothing: *Ex nihilo nihi fit*

There is nothing free in soil, it is the perception that soil is free that has caused its abuse and degradation. The law of return of soil management is based on the concept that 'nothing comes from nothing', implying that to get something from soil we have to add something to it (see Table 22.1).

22.2 Privilege of International Travel

Rattan Lal always felt fortunate and privileged in having enormous opportunities to travel around the world. His international travel began in December 1965, when at 21 he left India and travelled to Ohio, USA, where he studied for his PhD degree. By 2025, he had travelled to some 103 countries (Table 22.2). From 1965 to 1992, he travelled as an Indian citizen. From 1992 onwards, 5 years after migrating to the USA and becoming a U.S. citizen, he travelled as a U.S. citizen. While international travel is an opportunity for great

Table 22.1. Ecosystem services and law of nothingness. (Adapted from Lal, 2013)

The law or concept	Implications
Nothing is appropriated	There are always tradeoffs (give and take)
Nothing is permanent	Everything is in a dynamic equilibrium and a transient state
Nothing is absolute	All processes, properties and values are relative to a baseline
Nothing is a panacea	There is no silver bullet, there is a multitude/menu of options
Nothing is universal	Soil/site/region specifi city is an important consideration which cannot be overlooked
Nothing tangible is free	Under valuing a commodity leads to "Tragedy of the Commons"
Nothing is empty (vacuum) in nature	All space is occupied, pores in solid rock contain water or air and injecting something (liquid CO_2), fracking solutions can create shock waves
Nothing is given or for granted	It is the judicious use and management which produce goods and services
Nothing is a waste	Everything in nature has a use
Nothing comes from nothing (Ex nihilo nihil fit)	There is no such thing as a free lunch

Table 22.2. Countries visited by Rattan Lal from 1965 to 2025.

Region	Countries
Africa	Algeria, Benin, Burundi, Burkina Faso, Cameroon, Congo, Egypt, Ethiopia, Ghana, Ivory Coast, Kenya, Liberia, Libya, Madagascar, Malawi, Mali, Morocco, Niger, Nigeria, Rwanda, Senegal, Sierra Leone, South Africa, Tanzania, Togo, Tunisia, Uganda, Zaire, Zambia, Zimbabwe
Asia	Azerbaijan, Bangladesh, Cambodia, China, Dubai, India, Indonesia, Israel, Japan, Kuwait, Tajikistan, Lebanon, Malaysia, Nepal, Pakistan, Philippines, Qatar, Saudi Arabia, South Korea, Sri Lanka, Syria, Taiwan, Thailand, Turkey, UAE, Uzbekistan
South America	Argentina, Brazil, Colombia, Chile, Ecuador, Guyana, Peru, Venezuela
Central America	Costa Rica, Guatemala, Honduras, Nicaragua, Panama
North America	Canada, Mexico, USA
Caribbean	Dominican Republic, Haiti, Puerto Rico, Trinidad & Tobago
Europe	Austria, Belgium, Denmark, France, Georgia, Germany, Greece, Holland, Hungary, Iceland, Italy, Moldova, Norway, Poland, Portugal, Russia, Slovakia, Spain, Sweden, Switzerland, Turkey, Ukraine, UK, USSR
Pacific	Australia, Fiji, New Zealand

learning, some experiences are more rewarding and pleasant than others. However, as is the case for most of life's diverse experiences, the value of an experience is realized when the experience is unpleasant. Needless to say, international travel has been much easier while having a U.S. passport than an Indian passport. Travelling to any European country was a challenging task as an Indian national because of the strict visa requirements. Rattan learnt a lot about the language and culture of south-eastern and north Asian countries (Indonesia, Malaysia, Nepal, Bangladesh, Cambodia, Pakistan, China, Japan), central Asian countries (Uzbekistan, Kazakhstan, Azerbaijan, Tazikistan, etc.), Northern African countres (Egypt, Algeria, Morocco, Tunisia) and west Asian countries (Saudi Arabia, UAE, Lebanon, Syria), among others. While visiting Azerbaijan to attend COP29 in 2024, Rattan learnt that many words used in native Bacu language were the same as those in Urdu and Punjabi. The common words were *sahil* (beach or ocean front), *zaika* (taste), *durbar* (court), *shahi* (imperial), *bahar* (spring) and *safar* (journey). One of the reasons for these common words is that both Urdu and Punjabi contain words from Hindi, Farsi, Turkish, Arabic, etc. It is this synthesis which make Urdu and Punjabi rather interesting languages.

22.3 The Use of the Word 'Duniya'

The use of the word 'duniya' reminded Rattan of his favourite songs from some Bollywood movies. For example, the movie *Pyasa* had a song 'Tumhari hai tum hi sambhalo yeh Duniya, yeh Duniya agar mil bhi jai to kya hai' ('This world is yours and you can have it. Even if I get it, so what'). All one has to do is to google the word duniya and search for some other favourite Bollywood songs based on the word. Some examples of these are:

- 'Aisi Duniya' – a Hindi song by Padma Wadkar
- 'Duniya Bananewale' – a song from the movie Teesri Kasam by Mukesh
- 'Duniya Mein Sabse Hasin' – a song by Kishore Kumar
- 'Gol Gol Duniya' – a romantic song by R.D. Burman
- 'Meri Duniya...College Days' – a song by The Vintage Taan about the memories of school, college and friendship
- 'Roshan Tumhi Se Duniya' – a classic Hindi song by Mahipal and Geetanjali

Another popular word of Farsi origin is *waqat* or time. There are some popular movies based on this word and the related songs are among all-time favourites. The famous movie of the 1960s named *Waquat* had Bal Raj Sahni as the lead actor and told how he missed the country and his little daughter when he travelled to, and lived in, Kabul, Afghanistan.

22.4 Some Common Foods in South Asia and the Middle East

Rattan Lal learned a lot about common foods, habits and cultural traditions. For example, some of his favourite food, which he always thought was of Punjabi origin, turned out to be national dishes of other countries. The word *naan* for bread is commonly used in countries of Asia such as Afghanistan, Bangladesh, Bhutan, Iran, India, Nepal, Pakistan, Sri Lanka, the Middle East, Central Asia (Uzbekistan, Kazakhstan) and former Soviet Union countries.

Similar to *naan* from Central Asia, the word *jaleba* or *jalebi* and *halwa* are commonly used in Arabic and North African countries, *gulab jamun* in Turkey and *raso gulla* in Bangladesh. Rattan often heard arguments between Indians and Brazilians if the word 'annanas' (pineapple) wss given by Portuguese travellers from Brazil to India or some travelling from India to Brazil. He was once pleasantly surprised by a dialogue with a waiter about making a choice for breakfast in a hotel in Nairobi. The waiter asked, 'Would you care for Kenyan dishes like samosa, puri, pratha, idli, dosa, with sambar and papadum, or would you rather have western dishes such as omelette or scrambled eggs?' On his first ever trip to Brazil in 1975, Rattan wanted a cup of tea but did not know how to say it in Portuguese. He learnt later that the word for tea in Portuguese is 'chai'. Yes, indeed, the world is getting smaller with each generation of travellers, or even sooner. While visiting countryside in Morocco or Algeria, Rattan was often offered dried dates (*chuarra* in Punjabi or Urdu) with a cup of fresh milk and he remembered this custom in Punjab when an august guest came to visit a family.

22.5 Some Common Words in South Asian and Other Languages

There were also commonalities in some words used in Urdu and Punjabi with several countries in the Middle East and Central Asia. While visiting Azerbaijan, Rattan was pleasantly surprised with the use of some key words. For example, *duniya* (the world) was used as a slogan for COP29 (*yasil duniya*, or the green world). While travelling on a Turkish airline, Rattan found out that the name of an air hostess was *Bahar* (spring). Yes, indeed, the world is one family, and this we must be proud of and use it as the foundation for global peace and harmony.

22.6 Some Educational Experiences During International Travel

Rattan's travel was mostly economy class until about 2010. Thereafter, when he had the privilege to travel in business class, he experienced some basic nature of *yeh duniya* (this world) and of modern civilization. During earlier days, whenever Rattan had a rare business class ticket and he would be waiting in the specific line designated for that class, he was often questioned if he was standing in the right line. The first ever business-/first-class travel for Rattan was from Sydney (Australia) to Lagos (Nigeria) via New Delhi, when he joined IITA in December 1969. While boarding the Air India flight in New Delhi in December 1968 on his way to Beirut and then to Lagos, Rattan was stopped by the ground staff and asked to board the plane from the rear door, which was for economy-class passengers. Feeling somewhat embarrassed, he showed his boarding card to the staff who were also rather surprised to see that he had a first-class boarding card. This was the same experience he had when he was called for interview as a candidate for admission at the agricultural college in Ludhiana. The way he was dressed and his country-style haircut (short hair with a pony tail as a sign of being of a Brahmin family) caused the selection committee to think that he was not a candidate. He was asked to go out and send in the next candidate. He experienced similar embarrassing situations during his later travel but the frequency of such embarrassing incidences decreased as his hair turned grey and fell out, his face became wrinkled, and his back became hunched. The outer appearance give an impression that may be difficult to alter. Rattan eventually got the admission in Ludhiana, but it may have been easier if he was dressed more suitable for the occasion.

22.6.1 International vaccine certificate

During the annual leave in August 1978, eldest son Abhishek was two months old. Any traveller from West Africa to India had to have a medical certificate for the yellow fever among other vaccines. However, the medical doctor of IITA advised Rattan that a baby of less than 1 year cannot be given a yellow fever vaccination. Thus, he provided an official letter printed on the IITA letterhead paper with his signature and the ID of his medical certification. With the letter, Rattan's family (Sukh and three children) landed at the Indra Gandhi airport at about 11 pm. By the time their turn came at the medical check line it was about midnight. He presented all the certificates and the letter from the doctor. The medical officer said that the letter was not good enough because anyone (of any age) coming from West Africa must have the yellow fever vaccine. Rattan asked to see the man's supervisor, and all the officers assured him that he had two options: 1. return to Nigeria and come back with a valid certificate; or 2. go into quarantine for 10 days. He was further advised that only the two-month-old baby was under quarantine restriction; the rest of the family could go where they wished. Sukh said that she was not leaving the baby in the care of anyone. Priya (aged 4) and Bobby (aged 3) started protesting that they were not going anywhere without their mother and the baby. Thus, the medical supervisor said that the entire family could quarantine. The family was taken to the quarantine centre at about 2 am. While the children were asleep, they had to carry them inside one by one. When they put them on the beds, it was then that they realized that the facility was about 200m away from the active runway and the building and floor were shaking. There was also loud noise from the planes. It was a humbling experience to travel with young children. One third of their vacation was spent in quarantine. Despite the hardship, the quarantine staff were courteous, polite and played with children. There were no catering facilities and food had to be ordered from a *dhaba* in the nearby town. Finally, the children got used to the noisy environment and they would run to the window to see the planes taking off and landing.

22.6.2 The transit visa

As a transit passenger at a European airport during the 1970s and 1980s, Rattan often spent nights in the transit lounge as an economy-class passenger and without a pre-obtained visa to a specific European country. Under these circumstances, the only option was to deposit the passport with immigration officials and obtain a receipt. Upon returning to the airport the next day, Rattan would repossess the passport and continue his journey. However, an unexpectedly pleasant incident occurred at a Scandinavian country, where as a transit passenger he was getting mentally prepared to spend the night at the airport. An officer asked if he was not leaving the airport because he did not like the country. Rattan reluctantly explained to the officer that he had no visa to get out of the airport for the night. The officer asked him to show his ticket and passport. The kind officer consulted his colleagues and told Rattan that they had got a copy of his ticket and passport and was free to go to the city and come back in the morning to continue his journey.

22.6.3 A wet bag

Rattan Lal had a different experience when he arrived at New York in 1974 on a trip from Lagos to El Batan, Mexico (where the International Maize and Wheat Research Center, or CIMMYT, is located). He had the Mexican visa but not one for the USA. To his disappointment, his connecting flight from New York JFK airport to Mexico was cancelled, and he had to take another flight that departed from La Guardia airport. Thus, he had to leave JFK and be taken to another airport, and the problem was that he had no visa to enter the USA. Naturally, therefore, being a 'wet bag' (as officials called him), his Indian passport was taken by the authorities, and Rattan Lal was escorted to a separate room (detention centre) until arrangements were made for his transfer to La Guardia. Being an employee of the Rockefeller Foundation (RF) as a soil physicist at IITA, he waited until about 9 am and telephoned his contact at the RF in New York. Regrettably, the staff at RF explained to him that there was nothing to do but follow the rules of the U.S. immigration authorities. Because the flight to Mexico was in the late afternoon, the 'wet bag' had to stay for several hours at the specific facilities for unwelcome visitors. He was escorted to the rest room or any other facilities by an officer, and was eventually escorted to La Guardia. The police officer accompanying him to La Guardia was polite and courteous. To his surprise, the police officer escorted the 'wet bag' to a plane departing for Mexico. When Rattan Lal asked for the passport, police officer informed him that 'the passport is with the pilot, and will be given to him when the airplane has left U.S. air space'. The 'wet bag' and his Indian passport were reunited after the plane landed in Mexico.

22.7 Some Negative Experiences of Travel

Some frightening experiences also occurred to Rattan Lal both as an Indian and as a U.S. passport holder.

Once, at Rome airport during mid-1970s, he had to go to the airline office to rebook his flight. He was accompanied by a helpful female member of staff from the second floor to the checking counter on the ground floor. While she was leading him, an armed police officer asked her, in Italian, if she was safe, and she said 'yes'. The police officer watched Rattan he followed the airline official to the checking counter. To say the least, he was somewhat shaken by this experience of attracting negative attention from an armed police officer.

Once, he was travelling from IITA, Ibadan, to East Africa to participate in a conference. His rice-breeder colleague at IITA asked him if he would take the seeds of some varieties of rice, which needed to be tested in East Africa, and he was to give the package to an IITA staff member based there. Rattan was given an official letter certifying that he had some seeds for experimental purposes. Seeds of each variety, about 10 g, were contained in transparent sealed bags for easy examination by the quarantine authorities. On arrival at the airport, Rattan declared on the immigration forms that he had some seeds of rice. Having read the declaration forms, an officer asked Rattan to follow him. In a separate room, the officer asked him the amount of rice that he was importing. Rattan showed him the package (about 1 kg) containing seeds of hundreds of varieties. The officer was visibly annoyed and demanded to know if this was a some kind of joke, suggesting he should stop wasting his time.

When travelling to a developed country on an Indian passport, Rattan's immigration status, according to IITA, was that of a 'government official', which enabled travel to many countries without a specific visa. Yet immigration officers would often ask, 'On what page is the visa?' It would often be expected that visitors would leave a US$20 bill. Rattan would repeatedly state that he was a government official and did not need to show any visa.

In January 1972, Rattan arrived at Lagos airport with his wife. They had airfreighted a lot of wedding gifts as a part of their baggage. IITA was very generous to pay the freight charges. However, the immigration clearance took some time, and by the time they reached the baggage belt there were a couple of suitcases and nothing else. In them were some 20 oddly packed and nicely wrapped packages. Disappointed, Rattan filled in the forms for the missing goods and left the airport. He kept calling the airline to see if his packages had arrived and was told they had not. The travel agent advised him to go to the airport and look in the lost property room. The room was very disorganized and also hot and dusty, making it difficult to breathe. After several hours of searching, Rattan found one of his 20 packages. He returned the next day, spent another 6 hours looking, and found each of the gifts.

After Rattan Lal received the U.S. passport around the mid-1990s, because he had stated that his place of birth was Karyal, Pakistan, there were a lot of queries raised when he travelled to India and also when he arrived at any airport in the USA. What was a person who is a U.S. citizen and born in Pakistan doing in India? An easy way to solve this problem was to change the place of birth to 'Karyal, British India'. No one asked any questions after that.

A more serious incident occurred at Frankfurt airport in November 2024 when Rattan was in transit from Baku (where he attended the COP29) to Newark, USA. He was passing through the security check and his rollerboard and hand baggage were screened. His rollerboard was being closely scrutinized. From a distance of about 3–5 m (10–15 ft), he could see the computer screen on which several security officials were looking at Rattan and wondering who he was. The bag that was being scrutinized had a water-bottle that had been given to Rattan at COP29. On the X-ray, the bottle looked like a pipe. On top of the bottle was a portfolio (folder) with a writing pad and a small pocket containing a flash drive, and a rolled-up cable used to charge his cellphone. One of the officers walked up to Rattan and demanded to see his passport and asked him where he was coming from and what was his destination. Soon after the officer took possession of his passport. Meanwhile, an armed police officer appeared next to him. This was one of those times when Rattan Lal would recite his favourite Gayatri mantra and to stay as calm as he could. Finally, a Hindi-speaking staff arrived and started asking him questions in Hindi about what he did for a living, where he was travelling from, and where he was travelling to. Rattan informed him about the contents and that he packed the bag himself. Then the decision was made to open the bag. When Rattan Lal tried to touch the bag, someone shouted 'Don't touch It!', and this command startled him. The bag was finally opened by one of the officers. The water-bottle and portfolio were taken out and the charging cable and flash drive examined carefully. Being satisfied, the Hindi-speaking staff member returned the passport. Despite being stationary, Rattan was completely out of breath, sweat dripping from his forehead.

In March 2025, Rattan was travelling from Brasilia to Columbus via Panama and Newark. There was a plane change in Panama and Rattan Lal had to go through the security check. Being 80-plus, he had assumed that he did not have to take off his shoes. After taking of his belt, watch and wrist heart monitor, the officer asked him to take off his shoes. Rattan Lal responded that he was 80 and could not walk very well without his prescription shoes. This excuse was rejected. Walking barefoot across the security blockage is rather an uncomfortable experience that many elderly people endure because of inflexible rules.

22.8 Security of International Travel

Security for international travel is a high priority, and has been since the unfortunate event of 9/11. Nonetheless, there is a strong need for security officers to be courteous, polite, objective, rational and, above all, trusting, and to treat all passengers with respect. Mistrust and hasty judgements can lead to serious consequences. Similarly, passengers under suspicion must stay calm, co-operative, respectful and obey commands by officers. Since 9/11, their job has become more hazardous and a major challenge.

Despite all the risks and hazards, international travel is important in the 21st century, as it has been since the dawn of human civilization. Co-operation, trust and understanding is needed among passengers and the authorities to promote safety while strengthening trust, mutual respect and dignity.

22.9 The World Is One Family

Having visited over 100 countries, and experienced their culture and traditions, Rattan Lal has become a firm believer in the Sanskrit phrase 'Vasudhaiva Kutumbakam' (Sanskrit: वसुधैव कुटुम्बकम्). It appears in ancient Indian scriptures, such as the *Maha Upanishad*, the *Panchatantra* and the *Hitopdesha*. The phrase conveys a philosophy that encourages people to consider the entire universe as a family and to prioritize the collective good over the individual. A strong belief in this phrase is the long-term solution to securing world peace and harmony. Over time, and travelling internationally from 1965 to the present, Rattan has realized the importance of this ancient Indian philosophy.

In the context of 'The world is one family', the focus of young researchers must be on addressing issues of global significance, i.e. achieving climate change stabilization through adaptation and mitigation; advancing food and nutrition

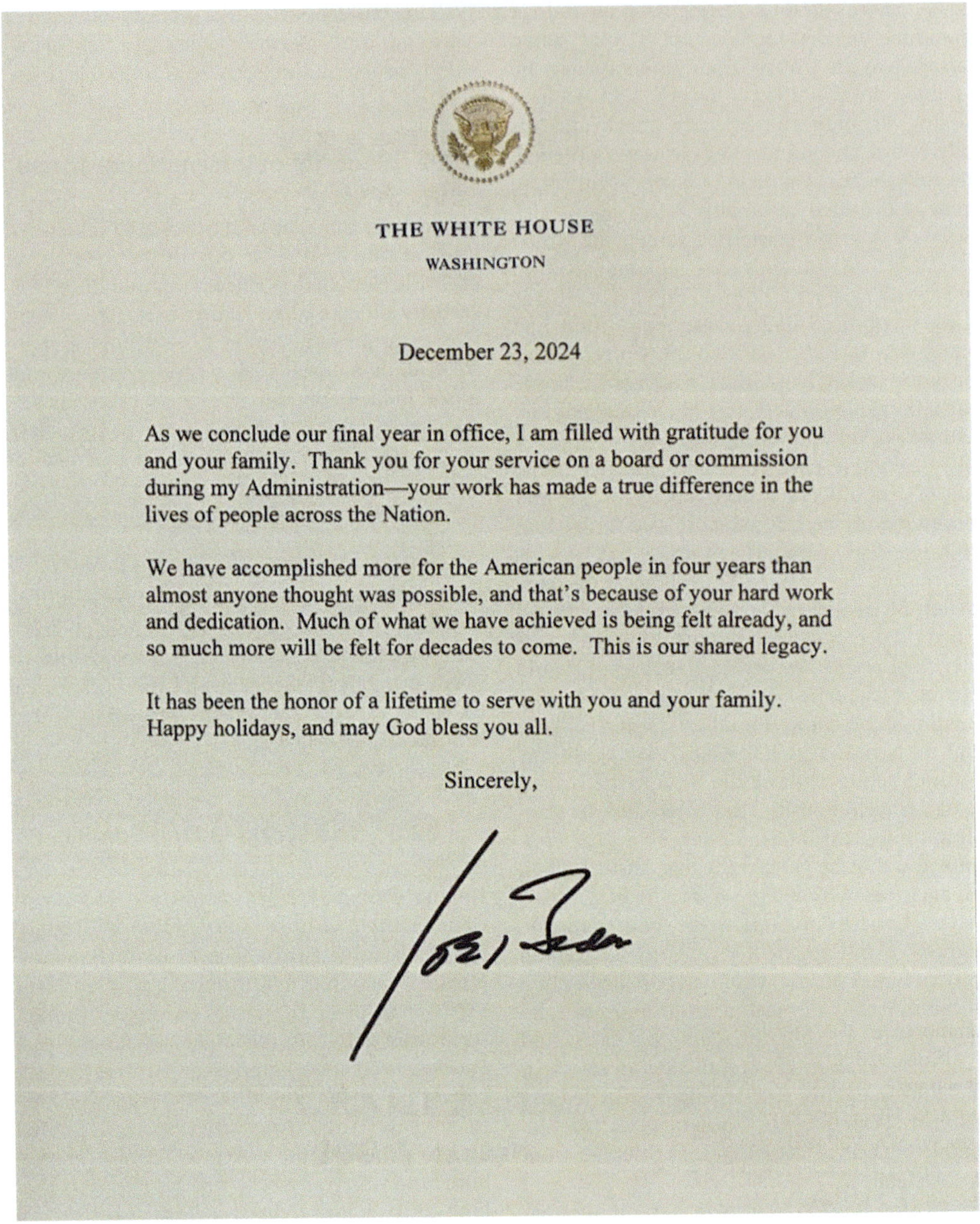

THE WHITE HOUSE

WASHINGTON

December 23, 2024

As we conclude our final year in office, I am filled with gratitude for you and your family. Thank you for your service on a board or commission during my Administration—your work has made a true difference in the lives of people across the Nation.

We have accomplished more for the American people in four years than almost anyone thought was possible, and that's because of your hard work and dedication. Much of what we have achieved is being felt already, and so much more will be felt for decades to come. This is our shared legacy.

It has been the honor of a lifetime to serve with you and your family. Happy holidays, and may God bless you all.

Sincerely,

Fig. 22.1. Letter from U.S. President Joe Biden on 23 December 2024. (Photo: Rattan Lal)

security; protecting soil functions; restoring degraded and depleted soils and desertified ecosystems; restoring above- and below-ground biodiversity; eliminating poverty and inequality; and promoting justice and peace. The strategy is to be positive, optimistic and hopeful. In Punjabi, there is a saying, 'Man ke hare har hai, aur man ke jeeta jeet' ('If the mind gives it then it is defeat, if the mind is victorious, it is the victory' (see Chapter 1). Furthermore, it is important not to focus on reputation but on doing good work for humanity and the planet with all determination and inner strength.

An example of 'The world is one family' is Rattan's appointment to BIFAD by former U.S. President Joe Biden (see Chapter 7.15). President Biden also shared a letter of service with Rattan on his last day in the White House (Fig. 22.1).

22.10 Love of Nature or *Kudrat* (Divine Power)

From childhood, Rattan Lal has been fascinated by *kudrat*. In Persian, the word is used to denote divine power or the supreme authority, or simply nature. I have always loved things that are moving upwards in their quest to take the world with them to new heights of glory and accomplishment. Some examples of these naturally powered or *kudrat*-driven things include:

The rising Sun
The shining Moon
The germinating seed
The growing plant
The blooming flower buds
The flying birds
The chirping of birds
The display of majestic feathers by a peacock
The flight of butterflies from one flower to another
The moving clouds
The thundering rainstorms
The towering snow-covered mountains
The flowing rivers
The magnificent waterfalls
The grazing cattle in the meadows
The nursing of a young calf by a cow
The tending of crops by farmers
The viewing of natural landscapes
The explaining by a youth what he/she wants to be

Above are some of Rattan's favourite things he loves most. He admires and is fascinated by nature. *Kudrat* is his source of strength, his motivation, and the source of his willpower and drive. As is stated in the *Upanishads* (Brihadaranyaka IV.4.5) (Easwaran, 1987):

You are what your deep, driving desire is.
As your desire is, so is your will.
As your will is, so is your deed.
As your deed is, so is your destiny.

Bibliography

Commoner, B. (1971) In: *The Closing Circle: Nature, Man, and Technology*. Alfred A. Knopf, New York. ISBN: 9780224006446.

Easwaran, E. (1987) *The Upnishads*. Nilgiri Press, Tomales, California. ISBN: 9781586380212.

Garret, G. (1968) Tragedy of the commons. *Science* 162, 1243–1248.

Gussow, J.D. (2024) The real organic podcast. The Organic Project. Available at: https://podcasts.apple.com/gb/podcast/real-organic-podcast/id1573111162 (accessed 15 June 2025).

Harding, G. and Baden, J. (eds) (1977) *Managing the Commons*. W.H. Freeman, San Francisco, California. ISBN.

Keats, J. (n.d.) A thing of beauty is a joyfor ever; its loveliness Increases, it will never pass into nothingness. Brainy Quotes. Available at: https://www.brainyquote.com (accessed 15 June 2025).

Kuhn, R. (2013) Levels of nothing. *Skeptic Magazine* 18(2), 34–37.

Lal, R. (2013). Soils and ecosystem services. In: *Ecosystem services and carbon sequestration in the biosphere* (pp. 11-38). Dordrecht: Springer Netherlands.

Mogahed, Y. (2015) *Reclaim Your Heart: Personal Insights on Breaking Free from Life's Shackles*. F.B. Publishing.

Muir, J. (1869) *My First Summer in the Sierra*. Better World Books.

The Holy Qur'an (2000) Wordsworth Editions Limited, Hertfordshire.

Index

Printed and bound by CPI Group (UK) Ltd, Croydon, CR0 4YY

13/09/2026

14961084-0001